高等职业教育新形态系列教材

机械专业英语

主　编　张文亭　王　滨

北京理工大学出版社
BEIJING INSTITUTE OF TECHNOLOGY PRESS

内 容 简 介

本书主要介绍机械专业英语的基础知识，对机械行业所涉及的专业领域知识采用英语进行了讲解，内容取材力求应用面广，并具有专业性强、词汇量和语言丰富、实用性强的特点。希望学生通过本书的学习，能达到提高阅读水平和应用专业英语的目的。

本书以培养学生专业英语能力为主要目标，全书精选 15 个单元：机械制图与建模、工程材料与热处理、电子电气应用、机械原理与机械设计、零件制作与普通加工、液压与气动回路设计、机械制造工艺及仿真、数控机床及应用、PLC 控制技术与应用、工业机器人及应用、机械产品质量检测与装配、逆向工程与 3D 打印、智能制造生产线及应用、机电设备保养与维护、生产现场优化与数字化管理。

本书可以作为高等院校机械设计制造及自动化、智能制造工程技术、数控技术、材料成型及控制工程、机械电子工程技术，高等职业教育专科机械设计与制造、数控技术、机械制造及自动化、模具设计与制造、机电设备技术、机电一体化技术等专业的专业英语用书，还可提供从事机械行业的工程技术人员参考。

图书在版编目(CIP)数据

机械专业英语 / 张文亭，王滨主编. -- 北京 : 北京理工大学出版社，2023. 12(2024. 1 重印)

ISBN 978-7-5763-3296-4

Ⅰ. ①机… Ⅱ. ①张… ②王… Ⅲ. ①机械工程-英语-教材 Ⅳ. ①TH

中国国家版本馆 CIP 数据核字(2024)第 014036 号

责任编辑：多海鹏　　**文案编辑**：辛丽莉
责任校对：刘亚男　　**责任印制**：李志强

出版发行 / 北京理工大学出版社有限责任公司
社　　址 / 北京市丰台区四合庄路 6 号
邮　　编 / 100070
电　　话 / (010) 68914026 (教材售后服务热线)
(010) 68944437 (课件资源服务热线)
网　　址 / http://www.bitpress.com.cn

版 印 次 / 2024 年 1 月第 1 版第 2 次印刷
印　　刷 / 三河市天利华印刷装订有限公司
开　　本 / 787 mm×1092 mm　1/16
印　　张 / 13
字　　数 / 282 千字
定　　价 / 45.00 元

前　言

本书主要介绍机械专业英语的基础知识，对机械行业所涉及的专业领域知识采用英语进行了讲解。内容取材力求应用面广，并具有专业性强、词汇量和语言丰富、实用性强的特点。希望学生通过本书的学习能达到提高阅读水平和应用专业英语的目的。

本书以培养学生专业英语能力为主要目标，全书精选15个单元：机械制图与建模、工程材料与热处理、电子电气应用、机械原理与机械设计、零件制作与普通加工、液压与气动回路设计、机械制造工艺及仿真、数控机床及应用、PLC控制技术与应用、工业机器人及应用、机械产品质量检测与装配、逆向工程与3D打印、智能制造生产线及应用、机电设备保养与维护、生产现场优化与数字化管理。

本书由张文亭、王滨主编。具体编写分工：陕西工业职业技术学院张文亭负责编写了单元1、单元4、单元6和单元11；陕西工业职业技术学院王滨负责编写了单元12、单元13、单元14和单元15；陕西工业职业技术学院王嘉明编写了单元2、单元5和单元7；陕西工业职业技术学院赵君编写了单元3、单元8、单元9和单元10。

本书可以作为高等院校机械设计制造及自动化、智能制造工程技术、数控技术、材料成形及控制工程、机械电子工程技术、高等职业教育专科机械设计与制造、数控技术、机械制造及自动化、模具设计与制造、机电设备技术、机电一体化技术等专业的专业英语用书，还可供从事机械行业的工程技术人员参考。

由于编者水平有限，书中难免有疏漏和不当之处，敬请读者多提宝贵意见，在此我们深表谢意。

编　者

目　　录

单元 1

机械制图与建模

【学习目标】

知识目标：

1. 熟悉机械制图与建模相关英语词汇。

2. 掌握与人交流绘制零件图、装配图及建模时常用的英语表达语句。

技能目标：

1. 具备机械制图与建模相关英语词汇寻读和跟读技巧，能运用英语检索相关信息。

2. 具备使用英语描述制图与建模相关信息和过程的能力。

素养目标：

1. 培养学生吃苦耐劳的工作精神。

2. 培养学生认真负责的工作态度和严谨细致的工作作风。

3. 培养学生制定并实施工作计划的能力、团队合作与交流的能力，以及良好的职业道德和职业情感，提高适应职业变化的能力。

Section Ⅰ Texts

Text 1

Engineering graphics is a cornerstone of engineering. The essence of engineering is design which requires graphics as the means of communication within the design process. Graphics serves as the common treaty for design of the manufacturing and construction processes.[1] Fig. 1-1-1 is an example of mechanical drawing.

Study of the fundamentals of engineering graphics is one key to your success as an engineer. Being able to describe an idea with a sketch is a prerequisite of the engineering profession. The ability to put forth a three-dimensional (3D) geometry in a form that can be communicated to other engineers, scientists, technicians, and nontechnical personnel is a valuable asset.[2] Of equal importance is knowing how to read and understand the graphics prepared by others.[3]

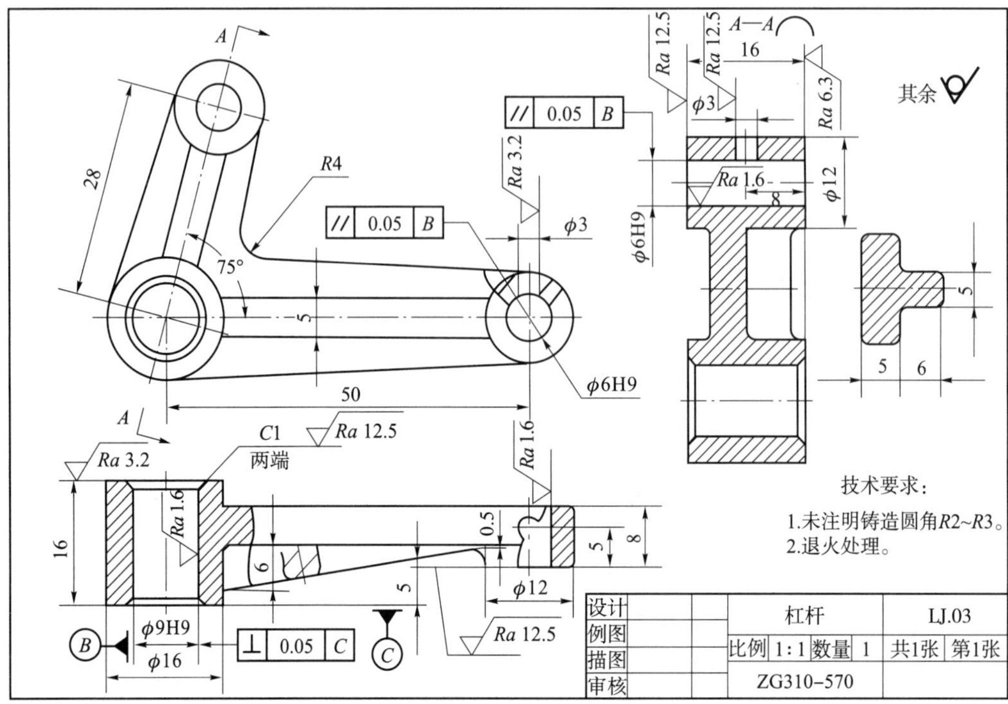

Fig. 1-1-1 Mechanical drawing

The ability to communicate is the key to success for a practicing engineer. Graphic communication, along with written and oral communication, constitutes an important part of a program of study in engineering. The fundamentals of the graphics language are universal in the industrialized world, an advantage not afforded by the written and spoken language. Thus, graphics may be said to be "a language for engineers".

The study of the graphics involves three aspects: terminology, skills and theory.

Engineering graphics is the area of engineering which involves the application of graphic principles in the development and conveyance of design concepts.

Engineering design is the systematic process by which a solution to a problem is created. Engineering graphics provides visual support, a basis for engineering analysis, and documentation for the design process.

Descriptive geometry is a set of principles which enable the geometry of an object to be identified and delineated by graphic means. It is the theory by which spatial (three-dimensional) problems involving angles, shapes, sizes, clearances, and intersections are solved with two-dimensional representation. Fig. 1-1-2 is an example of 2D (two-dimensional) drawing of end cover parts.

Computer graphics utilizes the digital computer to define, manipulate and display the devices, processes and systems for the purpose of the analysis, design and communication for engineering solutions.[4] See Fig. 1-1-3 computer 3D modeling.

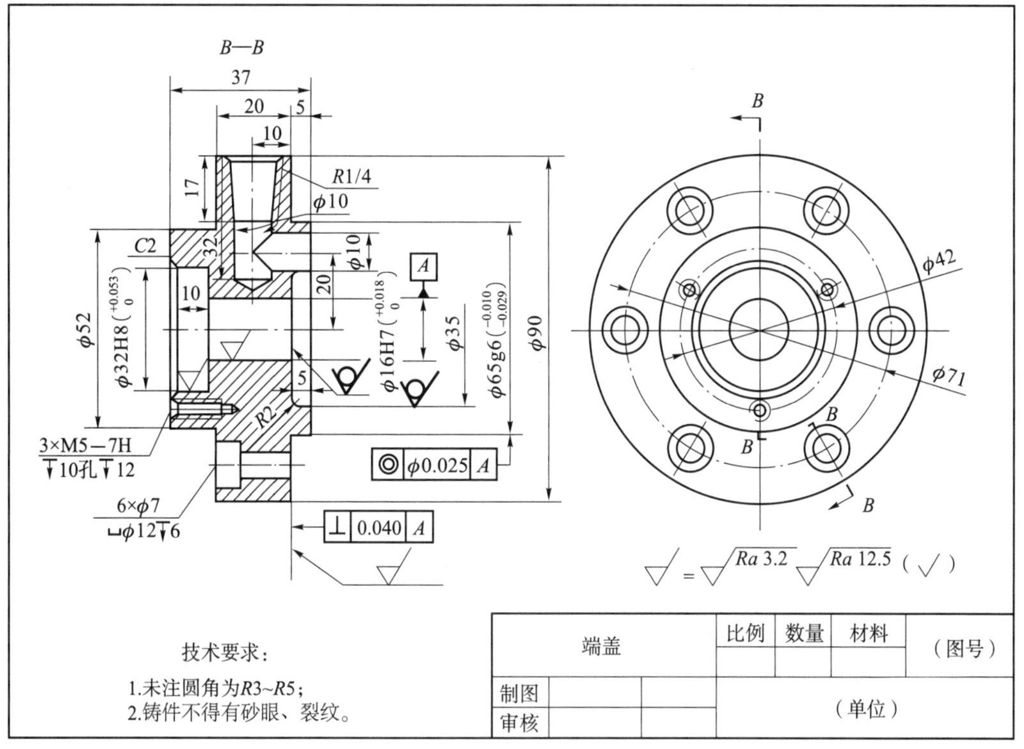

Fig. 1-1-2　2D drawing of end cover parts

Fig. 1-1-3　Computer 3D modeling

Geometric modeling is the representation of a concept, process, or system operation which is usually in a mathematical form and more specifically as an electronic database. Computer-based geometric modeling may conveniently be classified as wireframe, surface, or solid.

Engineering graphics is in a period of rapidly changing graphics technology. The traditional tools of the graphics, such as the T-square, compass and drafting machines, are being displaced by computer hardware and software. We are in an exciting era in which we will experience the transition from the scales, triangles and dividers to a computer keyboard and from the blueprints to databases.

The engineers, today, see the engineering drawing as a by-product of the CAD process. The control of the design-manufacture cycle is now the electronic database of the design. Changes are incorporated instantly in all aspects of the design. New product models can be quickly developed and oftentimes proved with the computer simulations, thus bypassing the prototype. If drawings are desired for manufacture or documentation, they may be quickly obtained from the database.

Today, the engineering student will study graphics from the standpoint of supporting the design process. The geometric modeling and analysis techniques which are mathematically based and practiced in the visualization of three-dimensional geometries will be the focus of intensive computer utilization.[5] In order to prepare the concepts for modeling and analysis, freehand techniques will be studied and practiced. The student will learn to produce and interpret multi-views and pictorials which are both via the sketches and computer techniques. Many of the graphics standards for the appropriate representation of the object. The student will learn to produce and interpret multi-views and pictorials which are both via the sketches and computer techniques. The features (sections, dimensioning, multi-views) will be studied.

Working in three dimensions with the computer, graphics will be produced easily in two-or three-dimensional modes depending upon the application.[6] Creating two-dimensional graphics such as *XY* plots and schematics will be accomplished with CAD software. Two-dimensional geometric primitives such as the circles and rectangles which serve as the generating geometry for the cylinders and prisms are a part of two-dimensional software. Special applications, for example, the dimensioning, generally utilize a two-dimensional view or series of two-dimensional views.

Three-dimensional geometric modeling involves wireframe, surface, and solid models. A wireframe model shows a series of the nodes connected by lines to form an object. The solid model is a total definition of an object which includes the knowledge of all boundary and internal points. From a solid model, a complete analysis of performance of the object can be performed on the computer with the appropriate software.

Computer graphics has become a powerful design tool which promises to enhance significantly the engineer's ability to be creative and innovative in the solution of the complex problems.

The information revolution is well under way. The rapid advancement of the electronic technology has changed the way we work and live. The language of graphics will continue to be a cornerstone of the communication for engineers and other technical persons. However, the changes we are seeing in the methods of transferring graphic, written and spoken material are astounding. These changes are improving the productivity of the industries and individuals as well as increasing the quality of the products and working environment. The requirements for the twenty-first century engineer will include a sound understanding of the fundamentals of graphics and the implementation of graphics to support the design process.

Exercises

1. Write T(True) or F(False) beside the following statements about the text.

(1) ________Graphics serves as the common treaty between the production and marketing processes.

(2) ________Engineering graphics is the area of engineering which involves the application of graphic principles in the development and conveyance of design concepts.

(3) ________Computer graphics utilizes the digital computer to define, manipulate and display the devices, processes and systems for the purpose of the production, communication and creation of the engineering solutions.

(4) ________Engineering graphics is in a period of not changing and revising graphics technology.

(5) ________The geometric modeling and analysis techniques which are mathematically based and practiced in the visualization of three-dimensional geometries will be the focus of intensive computer utilization.

(6) ________The student will learn to produce and interpret single-views and pictorials which are both via the photos and drawing techniques.

(7) ________Working in three dimensions with the computer, graphics will be produced easily in different modes depending upon the working purposes.

(8) ________Computer graphics has become a powerful design tool which promises to enhance significantly the engineer's ability to be creative and innovative in the solution of the complex problems.

2. Choose the best answer.

(1) In a normal view of line, the ___①___ length of a line ___②___ measured.

① A. Shorter B. longer C. equal D. truth

② A. should be B. can be C. is D. was

(2) It is common to ___①___ the front, side, and back views ___②___ elevations and to the ___③___ vies as ___④___ views.

① A. refer to B. as C. was D. should

② A. is B. as C. was D. should

③ A. right and left B. top and bottom C. front and back

④ A. normal B. principal C. plane D. elevations

(3) A normal view of a line is a ___①___ projection of the line onto a viewing plane ___②___ to the line.

① A. vertical B. sloping C. inclined D. horizontal

② A. cross B. parallel C. level D. perpendicular

(4) ___①___, only ___②___ of the three dimensions of width, height and depth can

be measured.

① A. Generally　　B. So that　　C. Therefore　　D. So

② A. one　　B. two　　C. three　　D. zero

(5) In the horizontal auxiliary view (auxiliary elevation), the ____①____ of an object can be measured. In a ____②____ auxiliary view, the object's depth can be measured.

① A. length　　B. width　　C. height　　D. depth

② A. top　　B. bottom　　C. back　　D. front

3. Fill in the missing words according to the text.

(1) The ability to put forth a ________ in a form that can be communicated to other engineers, scientists, technicians, and nontechnical personnel is a________.

(2) Engineering graphics is the area of engineering which involves the ________ in the development and ________.

(3) If drawings are desired for ________, they may be quickly obtained from ________.

(4) Two-dimensional geometric primitives such as the circles and rectangles which serve as the ________ for the cylinders and prisms are a part of two-dimensional software. Special applications, for example, the dimensioning, generally utilize ________ or series of ________.

(5) Computer graphics has become________which promises to enhance significantly the engineer's ability to be________ in the solution of the complex problems.

4. Please answer the following questions in Chinese according to what you have learned.

(1) 什么是正视图？正视图中任何图形的真实尺寸和形状都是可测量的吗？为什么？

(2) 什么是基视图？可用一个基视图完整地表示一个对称的物件吗？请解释原因。

(3) 在什么情况下需要使用辅助视图？为什么说长、宽、高三维度尺寸中仅有一个可以测量？

(4) 什么是剖面图？

(5) 说明工程图的用途。

Text 2

Then, if we have a computer, how can we control the machines? We should use some software. CAD and CAM are two terms which means computer-aided design and computer-aided manufacturing. It is the technology concerned with the use of digital computers to perform certain functions in design and production. This technology is moving in the direction of greater integration of design and manufacturing, two activities which have traditionally been treated as distinct and separate functions in a production firm.[7] Ultimately, computer-aided design and computer-aided manufacturing will provide the technology base for the computer-integrated factory of the future.

Computer-aided design can be defined as the use of computer systems to assist in the creation, modification, analysis, or optimization of a design.[8] The computer systems consist of the hardware and software to perform the specialized design functions required by the particular user firm. The CAD hardware typically includes the computer, one or more graphics display terminals, keyboards, and other peripheral equipment. The CAD software consists of the computer programs which implement computer graphics on the system plus the application programs to facilitate the engineering functions of the user company. Examples of these application programs include stress-strain analysis of components, dynamic response of mechanisms, heat-transfer calculations, and numerical control part programming (Fig. 1-1-4). The collection of the application programs will vary from one user firm to the next because their production lines, manufacturing processes and customer markets are different. These factors give a rise to the differences in CAD system requirements.

Fig. 1-1-4 Computer use

Computer-aided manufacturing can be defined as the use of computer systems to plan, manage and control the operations of a manufacturing plant through either direct or indirect computer interface with the plant's production resources.[9] As indicated by the definition, the applications of computer-aided manufacturing fall into two broad categories.

(1) Computer monitoring and control. These are the direct applications in which the computer is connected directly to the manufacturing process for the purpose of monitoring or controlling the process.

(2) Manufacturing support applications. These are the indirect applications in which the computer is used in support of the production operations in the plant, but there is no direct interface between the computer and the manufacturing process.

In addition to the applications involving a direct computer-process interface for the purpose of process monitoring and control, computer-aided manufacturing also includes indirect applications in which the computer serves a support role in the manufacturing operations of the plant. In these applications, the computer is not linked directly to the manufacturing process. Instead, the computer is used "off-line" to provide plans, schedules, forecasts, instructions, and information by which the firm's production resources can be managed more effectively (Fig. 1-1-5). The form of the relationship between the computer and the process is represented symbolically in the figure given below. Dashed lines are used to indicate that the communication and control link is an off-line connection, with human beings often required to consummate the interface. However, human beings are presently required in the application either to provide input to the computer programs or to interpret the computer output and implement the required action (Fig. 1-1-6).

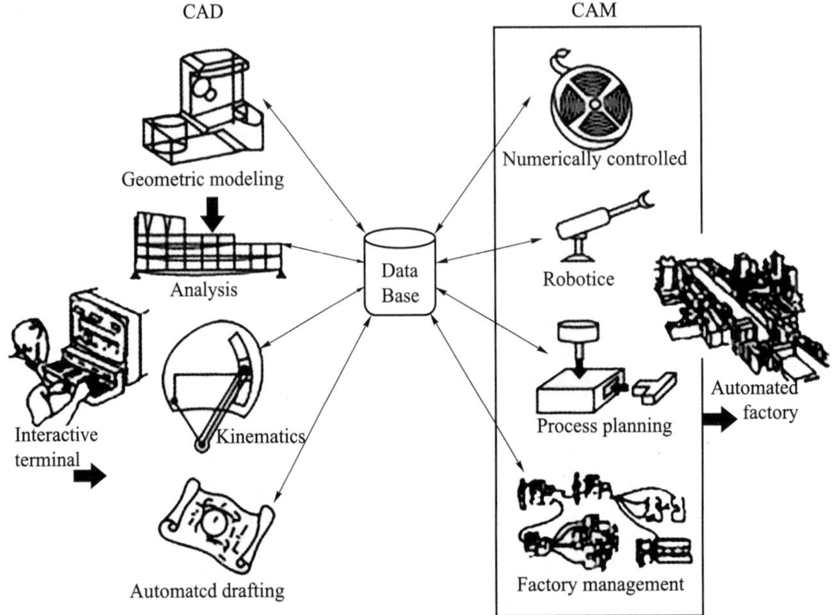

Fig. 1-1-5 The Conceptual CAD/CAM system

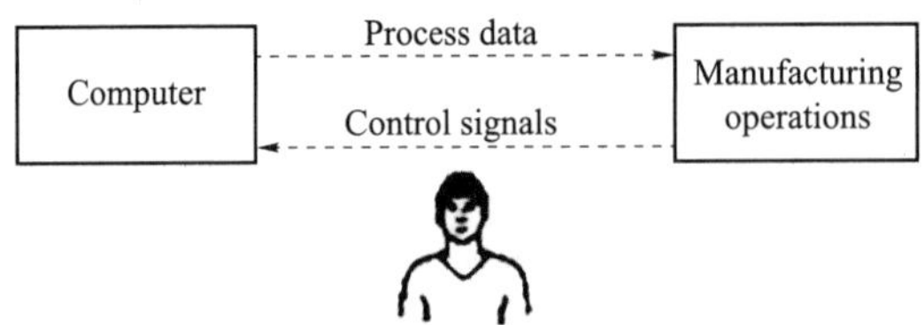

Fig. 1-1-6 CAM for manufacturing support

Section Ⅱ New Words and Phrases

graphics	*n.* 制图学，图形学，图解
cornerstone	*n.* 基础，基石
essence	*n.* 本质，要素，精华
means of communication	交流工具（手段），通信工具（手段）
fundamental	*n.* 基本原理 *adj.* 基本的
prerequisite	*n.* 先决条件 *adj.* 首要的；必备的
put forth	提出，发表，拿出
nontechnical	*adj.* 非技术性的
practicing	*adj.* 开业的，从业的，在工作的

industrialized	*adj*. 工业化的
conveyance	*n*. 运输，运输工具，运送，传送
descriptive geometry	画法几何
delineate	*v*. 描绘，描述
geometric modeling	几何建模，几何模型建立，形状模型化
T-square	*n*. 丁字尺
compass	*n*. 圆规
drafting	*v*. 制图，起草
scale	*n*. 比例尺，刻度尺
divider	*n*. 分规，两脚规
blueprint	*n*. 蓝图，设计图
database	*n*. 数据库
by-product	*n*. 副产品，副产物
prototype	*n*. 原型，样机，样品
geometric modeling	几何建模
visualization	*n*. 可视化，显形
freehand	*adj*. 徒手画的
pictorial	*adj*. 用图表示的，图解的
section	*n*. 截图
dimensioning	*v*. 标注尺寸
plot	*n*. 曲线，图形
schematic	*adj*. 示意的，简略的 *n*. 简图
node	*n*. 结点，交点，中心点
primitive	*adj*. 原始的，基本的 *n*. 基元，图元
rectangle	*n*. 长方形，矩形
prism	*n*. 棱柱
wireframe model	*n*. 线框模型
surface model	*n*. 曲面模型
solid model	*n*. 实体模型
innovative	*adj*. 创新的，革新的
astound	*v*. 使……大吃一惊，令人震惊
sound	*adj*. 可靠的，合理的 *adv*. 彻底地，充分地
computer-aided design (CAD)	计算机辅助设计
computer-aided manufacturing (CAM)	计算机辅助制造
integration	*n*. 结合，集成
traditionally	*adj*. 传统地
distinct	*adj*. 明显的，独特的

separate	*adj.* 分开的，单独的
ultimately	*adv.* 最后，根本上
computer-integrated	计算机集成的
optimization	*n.* 最优化
terminal	*n.* 终端，终端机
peripheral	*adj.* 外围的　*n.* 外围设备
graphics	*adj.* 制图法，制图学，图形
stress-strain	应力-应变
interface	*n.* 接口（外部设备用），界面
category	*n.* 种类，类别
monitor	*v.* 监控

Section Ⅲ　Notes to Complex Sentences

[1] Graphics serves as the common treaty between design and the manufacturing and construction processes.

图样好比是设计与加工制作过程中的共同规定。

[2] The ability to put forth a three-dimensional geometry in a form that can be communicated to other engineers, scientists, technicians and nontechnical personnel is a valuable asset.

作出三维几何图，让工程师、科学家、技术人员和非技术人员能够看懂，这种能力就是宝贵的财富。

[3] Of equal importance is knowing how to read and understand the graphics prepared by others.

能够看懂其他人制作的图纸也同样重要。这是一个倒装句，主语是 knowing how…，of equal importance 作表语，意为“具有同样重要性”。

[4] Computer graphics utilizes the digital computer to define, manipulate and display devices, processes and systems for the purpose of the analysis, design and communication of engineering solutions.

计算机绘图利用数字计算机来定义、操作和显示设备、过程及系统，是分析、设计和交流工程问题的途径。

本句是一个简单句，谓语是 define, manipulate and display 这三个并列动词，它们共同的宾语为 devices, processes and systems.

[5] The geometric modeling and analysis techniques which are mathematically based and practiced in visualization of three-dimensional geometries will be the focus of intensive computer utilization.

以数学为基础并通过三维几何图形制作的几何建模和分析技术，将是计算机应用的重要方面。

[6] Working in three dimensions with the computer, graphics will be produced easily in two- or three-dimensional models depending upon the application.

用计算机制作三维图形，即依据实际，通过二维或三维模型来生成图形。

[7] This technology is moving in the direction of greater integration of design and manufacturing, two activities which have traditionally been treated as distinct and separate functions in a production firm.

句中，two activities 指 design and manufacturing，which 引导定语从句修饰 two activities。

这种技术正向着设计与制造更加紧密结合的方向发展。在制造公司，设计与制造通常被认为是明显分离的两个方面。

[8] Computer-aided design can be defined as the use of computer systems to assist in the creation, modification, analysis, or optimization of a design.

计算机辅助设计是指用计算机系统来协助一个设计方案的形成、修改、分析及优化。

[9] Computer-aided manufacturing can be defined as the use of computer systems to plan, manage, and control the operations of a manufacturing plant through either direct or indirect computer interface with the plant's production resources.

计算机辅助制造指的是通过计算机直接或间接与车间生产设备连接并使用计算机系统来计划、管理和控制制造车间的操作。

Section Ⅳ Supplementary Reading

1. An Overview of Engineering Drawing

Engineering drawings are often referred to as "blueprints" or "bluelines". However, the terms are rapidly becoming an anachronism, since most copies of engineering drawings that were formerly made using a chemical-printing process that yielded graphics on blue-colored paper or, alternatively, of blue-lines on white paper, have been superseded by more modern reproduction processes that yield black or multi-color lines on white paper. The more generic term "print" is now in common usage in the U. S. to mean any paper copy of an engineering drawing.

Engineering drawings can now be produced by using computer technology. Drawings are extracted from three dimensional computer models and can be printed as two dimensional drawings on various media formats (color or monochrome). Engineered computer models can also be printed in the three dimensional form using special 3D printers.

The process of producing engineering drawings and the skill of producing them are often referred to be as the technical drawing, although technical drawings are also required for disciplines that would not ordinarily be thought of being as parts of the engineering.

Drawings convey the following critical information.

(1) Geometry — the shape of the object; represented as views; how the object will look when it is viewed from various standard directions, such as front, top, side, etc.

(2) Dimensions — the size of the object is captured in the accepted units.

(3) Tolerances — the allowable variations for each dimension.

(4) Material — represents what the item is made of.

(5) Finish — specifies the surface quality of the item, functional or cosmetic. For example, a mass-marketed product usually requires a much higher surface quality than, say, a component that goes inside industrial machinery.

2. Creating Drawings with AutoCAD

This text introduces AutoCAD for Windows, a software widely used in industry and production.

When you first start AutoCAD, an empty and unnamed drawing is open. You can begin working with this drawing immediately.

To create a new drawing, choose Start from Scratch.

(1) From the File menu, click New.

(2) In the Create New Drawing dialog box, choose Start from Scratch (Fig. 1-4-1).

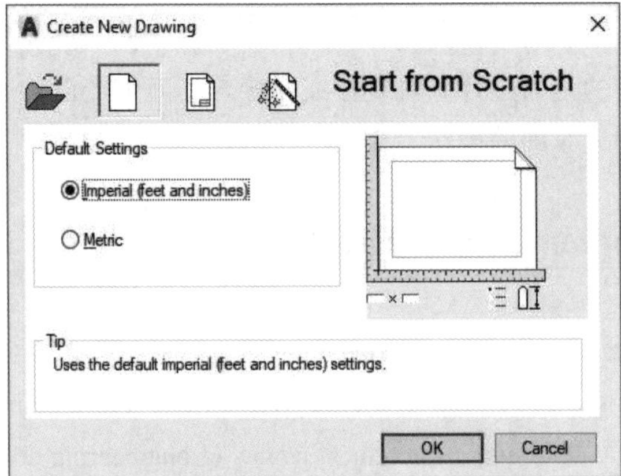

Fig. 1-4-1 Create new drawing dialog box

(3) In the Select Default setting, select English or Metric, and then click OK. The drawing opens with the default AutoCAD settings.

(4) From the File menu, click Save As.

(5) In the Save Drawing As dialog box under File Name, enter a name for the drawing and click OK.

The drawing extension (.dwg) is automatically appended to the file name.

Working with layers

An AutoCAD drawing can have layers, which is similar to the transparent overlays drafters used to create pencil drawings. For example, in a drawing of a house, you can have the wall lines on one layer, the electrical wiring on a second layer, and plumbing on a third. It's helpful to use different colors for the different layers, so you can tell them apart.

Follow the following steps to create a new layer.

(1) From the format menu, click Layer.

(2) In the Layer Properties dialog box (Fig. 1-4-2), choose New. A layer called Layer 1 is displayed.

(3) Click Layer 1 and enter a new layer name.

The layer name can include up to 31 characters. Layer name can contain letters, digits and the special characters dollar sign ($), hyphen (-), and underscore (_). Layer names cannot include blank spaces.

(4) To change a layer's color, select the layer and click its Color icon. In the Select Color dialog box, select a color and click OK.

(5) To change a layer's Line type, select the layer and click its Line type icon. In the Select Line type dialog box, select a Line type.

(6) Choose OK to exit from each dialog box.

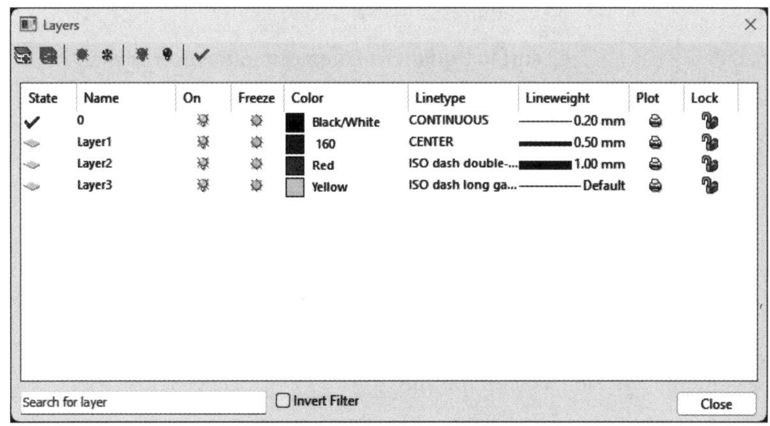

Fig. 1-4-2 Layer and line type properties dialog box

Exercises

1. Write T(True) or F(False) beside the following statements about the text.

(1) __________ CAD and CAM are two terms which mean computer-aided design and computer-aided manufacturing.

(2) __________ Computer-aided design can be defined as the use of computer systems to assist in the creation, modification, analysis, or optimization of a product.

(3) __________ The CAD software consists of the computer programs which implement computer programming on the system plus the application programs to make the management more effective.

(4) __________ The collection of the application programs will vary from one user firm to the next because their production lines, manufacturing processes and customer markets are different. These factors give a rise to the differences in CAD system requirements.

(5) __________ Computer-aided manufacturing can be defined as the use of computer systems to plan, manage and control the operations of a manufacturing plant through direct computer software with the plant's production people.

(6) __________ Computer monitoring and control are the direct applications in which the computer is connected directly to the manufacturing process for the purpose of monitoring or controlling the process.

(7) __________ In these applications, the computer is linked indirectly to the management process.

(8) __________ However, human beings aren't presently required in the application either to provide the computer programs with the computer output and implement the required action.

2. Match the following terms to appropriate definition or expression.

(1) CAD	a. computer aided manufacturing
(2) CAM	b. computer aided design
(3) geometry	c. surface quality of item, functional or cosmetic
(4) dimension	d. the shape of the object
(5) product quality	e. the size of the object is captured in the accepted units

3. Fill in the missing words according to the text.

(1) CAD and CAM are two terms which mean__________ and__________.

(2) It is the technology concerned with the use of__________to perform certain functions in __________.

(3) The computer systems consist of the __________ and __________ to perform the specialized design functions required by the particular user firm.

(4) Drawings are extracted from__________ and can be printed as__________ on various media formats (color or monochrome).

4. Translate the following paragraphs into Chinese.

(1) Engineering drawings can now be produced using computer technology. Drawings are extracted from three-dimensional computer models and can be printed as two-dimensional drawings on various media formats (color or monochrome). Engineered computer models can also be printed in three dimensional form using special 3D printers.

(2) The process of producing engineering drawings and the skill of producing them are often referred to as technical drawing, although technical drawings are also required for disciplines that would not ordinarily be thought of as parts of engineering.

(3) Engineering drawings can now be produced by using computer technology. Drawings are extracted from three dimensional computer models and can be printed as two dimensional drawings on various media formats (color or monochrome). Engineered computer models can also be printed in the three dimensional form using special 3D printers.

单元评价

通过本单元课文阅读，词汇、习题的练习，以及扩展阅读的学习，使学生掌握机械制图与建模相关的英语词汇及常用表达语句，并具备运用英语检索机械制图与建模相关信息及使用英语描述制图与建模相关内容的能力，培养学生吃苦耐劳的工作精神、认真负责的工作态度及团队合作与交流的能力。

单元 2

工程材料与热处理

【学习目标】

知识目标：

1. 熟悉工程材料与热处理相关英语词汇。
2. 掌握工程材料热处理的原理、作用及工艺编制常用的表达语句。

技能目标：

1. 具备工程材料与热处理相关英语词汇寻读和跟读技巧，能够运用英语检索相关信息。
2. 具备使用英语描述工程材料与热处理相关信息的能力。

素养目标：

1. 培养学生创新的思维能力以及严谨求实的学习态度。
2. 培养学生勤于动手、勤于思考、善于总结的良好习惯。
3. 培养学生爱岗敬业、精益求精、不断创新的工匠精神。

Section Ⅰ Texts

1. Introduction

Common engineering materials are normally classified as the metals and non-metals. Metals may conveniently be divided into ferrous and non-ferrous metals.

Important ferrous metals for the present purpose are:

(1) cast iron (Fig. 2-1-1);

Fig. 2-1-1 Cast iron

(2) wrought iron;

(3) steel (Fig. 2-1-2).

Fig. 2-1-2　Steel

Some of the important non-ferrous metals used in the engineering design are:

(1) light metal group such as the aluminium and its alloys.

(2) copper based alloys such as brass (Cu-Zn), bronze (Cu-Sn).

(3) white metal group such as nickel, silver, zinc, etc.

Fig. 2-1-2 shows chemical composition table of metal material.

牌号	Chemical composition/%								
	C	Mn	Si	P≤	S≤	Cr	Ni	Mo	Cu≤
45	0.42~0.50	0.50~0.80	0.17~0.37	0.035	0.035	0.25	0.25		0.25
16Mn	0.12~0.20	1.20~1.60	0.20~0.60	0.035	0.035	≤0.30	≤0.30		0.25
20MnMo	0.17~0.23	1.10~1.40	0.17~0.37	0.035	0.035	≤0.30	≤0.30	0.20~0.35	0.25
40Cr	0.37~0.44	0.50~0.80	0.17~0.37	0.035	0.035	0.80~1.10	≤0.30		0.30
35CrMo	0.32~0.40	0.40~0.70	0.17~0.37	0.035	0.035	0.80~1.10	≤0.30	0.15~0.25	0.30
5CrMnMo	0.50~0.60	1.20~1.60	0.25~0.60	0.030	0.030	0.60~0.90		0.15~0.30	0.30
34CrNi3Mo	0.30~0.40	0.50~0.80	0.17~0.37	0.030	0.035	0.70~1.10	2.75~3.25	0.25~0.40	
40CrNiMoA	0.37~0.44	0.0~0.80	0.17~0.37	0.035	0.035	0.60~0.90	1.25~1.65	0.15~0.25	0.30
0Cr18Ni9Ti(304)	≤0.08	≤2.00	≤1.00	0.035	0.030	17.0~19.0	8.00~11.0		Ti:5(C-0.02)~0.80
0Cr18Ni12MoTi(316)	≤0.12	≤2.00	≤1.00	0.035	0.030	16.0~19.0	11.0~14.0	1.80~2.50	Ti:5(C-0.02)~0.80
0Cr13Ni6Mo2Re	≤0.06	0.60~0.90	≤0.80	0.035	0.035	12.0~14.0	5.00~6.50	2.00~2.50	Re=0.20
0Cr13	≤0.08	≤0.60	≤0.60	0.035	0.030	12.0~14.0			
0Cr25Ni5Mo2	≤0.07	≤1.00	≤1.00	0.040	0.030	23.5~25.5	5.00~6.00	1.50~2.50	N≤0.05

Fig. 2-1-3　Chemical composition table of metal materials

2. Ferrous Materials

(1) Cast iron. It is an alloy of iron, carbon and silicon and it is hard and brittle.

Carbon content may be within 1.7%-3% and carbon may be present as free carbon or iron carbide Fe_3C.

(2) Wrought iron(Fig. 2-1-4). This is a very pure iron where the iron content is of the order of 99.5%.[1] It is produced by re-melting pig iron and some small amount of silicon, sulphur, or phosphorus may be present.[2] It is tough, malleable and ductile and can easily be forged or welded. It cannot however take a sudden shock. Chains, crane hooks, railway couplings and such other components may be made of this iron.

Fig. 2-1-4 Wrought iron

(3) Steel. This is by far the most important engineering material and there is an enormous variety of the steel to meet a wide variety of engineering requirements. Steel is basically an alloy of iron and carbon in which the carbon content can be less than 1.7% and carbon is present in the form of iron carbide to impart hardness and strength. Two main categories of steel are plain carbon steel and alloy steel.

① Plain carbon steel. The properties of plain carbon steel depend mainly on the carbon percentages and other alloying elements are not usually present in 0. 5%-1% such as 0. 5% Si or 1% Mn, etc.[3]

② Alloy steel(Fig. 2-1-5). These are steels in which elements other than carbon are added in sufficient quantities to impart desired properties, such as wear resistance, corrosion resistance, electric or magnetic properties.[4] Chief alloying elements added are usually nickel for strength and toughness, chromium for hardness and strength, tungsten for hardness at elevated temperature, vanadium for tensile strength, manganese for high strength in hot rolled and heat treated condition, silicon for high elastic limit.

Fig. 2-1-5 Alloy steel

3. Non-ferrous Metals

Metals containing elements other than iron as their chief constituents are usually referred to as non-ferrous metals. There is a wide variety of non-ferrous metals in practice. However, only a few exemplary ones are discussed below.

(1) Aluminium(Fig. 2-1-6). This is the white metal produced from Alumina. In its pure state it is weak and soft but addition of small amounts of Cu, Mn, Si and Mg makes it hard and strong. It is also corrosion resistant, low mass and non-toxic.

Fig. 2-1-6 Aluminium

(2) Magnalium(Fig. 2-1-7). This is an aluminium alloy with 2%-10% Mg. It also contains 1.75% Cu. Due to its light mass and good strength, it is used for aircraft and automobile components.

Fig. 2-1-7 Magnalium

(3) Copper alloys (Fig. 2-1-8). Copper is one of the most widely used non-ferrous metals in industry. It is soft, malleable and ductile and is a good conductor of heat and electricity. The following two important copper alloys are widely used in practice.

Fig. 2-1-8 Magnalium

Brass (Cu-Zn alloy) (Fig. 2-1-9)—It is fundamentally a binary alloy with Zn up to 50%. Brass is highly corrosion resistant, easily machinable and therefore a good bearing material.

Fig. 2-1-9 Brass

Bronze (Cu-Sn alloy) (Fig. 2-1-10)—This is mainly a copper-tin alloy where tin percentage may vary between 5% to 25%. It provides hardness but tin content also oxidizes resulting in brittleness.[5] It was originally made for casting guns but used now for boiler fittings, bushes, glands and other such uses.

Fig. 2-1-10 Bronze

4. Non-metals

Non-metals are also used in the engineering practice due to, principally, their low cost, flexibility and resistance to the heat and electricity. Though there are many suitable non-metals, plastics is the most important from the design point of view. They are synthetic materials which can be moulded into the desired shapes under pressure with or without the application of the heat. They are now extensively used in various industrial applications for their corrosion resistance, dimensional stability and relatively low cost.

Section Ⅱ New Words and Phrases

classify *vt.* 分类

conveniently *adv.* 方便地

ferrous	*adj.* 含铁的，铁的
cast iron	铸铁
wrought	*vt.* 锻造
wrought iron	可锻铸铁
alloy	*n.* 合金
nickel	*n.* 镍
zinc	*n.* 锌
silicon	*n.* 硅
brittle	*adj.* 脆的
carbide	*n.* 碳化物
iron carbide	*n.* 碳化铁
pig iron	生铁
sulphur	*n.* 硫
phosphorus	*n.* 磷
malleable	*adj.* 可塑的
crane	*n.* 起重机
hook	*n.* 钩子
coupling	*n.* 配件
ductile	*adj.* 可延展的；韧的
forge	*v.* 锻造；伪造
by far	到目前为止
hardness and strength	硬度和强度
category	*n.* 类别
plain carbon steel	（纯）碳钢
impart	*vt.* 添加；传给
corrosion resistance	抗腐蚀性
hot roll	热轧
elastic limit	弹性极限
toughness	*n.* 韧性
tensile	*adj.* 拉伸的
exemplary	*adj.* 举例的
constituent	*n.* 成分
tungsten	*n.* 钨
chromium	*n.* 铬
vanadium	*n.* 钒
manganese	*n.* 锰
non-toxic	*adj.* 无毒的

binary alloy	二元合金
brass	*n.* 黄铜
bearing	*n.* 轴承
bronze	*n.* 青铜
boiler fittings	锅炉配件
oxidize	*v.* 氧化
gland	*n.* 密封盖，端盖
flexibility	*n.* 灵活性
synthetic	*adj.* 合成的
mould mold	*n.* 模具 *v.* 模塑
extensively	*adv.* 广泛地
dimensional stability	尺寸稳定性

Section Ⅲ Notes to Complex Sentences

[1] This is a very pure iron where the iron content is of the order of 99.5%.

这是一种纯铁，其铁含量约占 99.5%。

to be of the order 意为“在……数量级”。

[2] It is produced by re-melting pig iron and some small amount of silicon, sulphur or phosphorus may be present.

它是由生铁重新熔化而成的，含有少量硅和硫，或少量磷。

or 引导的从句表示另一种选择，与主句平行。

[3] The properties of plain carbon steel depend mainly on the carbon percentages and other alloying elements are not usually present in more than 0.5% - 1%, such as 0.5% Si or 1% Mn, etc.

纯碳钢的性能主要取决于含碳百分比，其他合金元素通常不到 0.5%-1%，如 0.5%的硅或者 1%的锰。

and other alloying elements...由 and 连接的两个并列复合句。

[4] These are steels in which the elements other than the carbon are added in sufficient quantities to impart the desired properties, such as the wear resistance, the corrosion resistance, electric or magnetic properties.

这种钢添加了除碳以外的足够数量的其他元素，用以提供期望的性能，如抗磨损性、抗腐蚀性、电磁特性等。

such as...修饰 properties。

[5] It provides hardness but tin content also oxidizes resulting in brittleness.

它提供硬度，但锡会氧化造成脆弱。

Section Ⅳ Exercises

Please find answers from the text to the following questions.

Q1: Classify common engineering materials.

Q2: What are the advantages of malleable cast iron over white or gray cast iron?

Q3: A standard alloy steel used for making engineering components is 20Cr18Ni2. State the composition of the steel.

Q4: How is the plain carbon steel designated?

Q5: Name two important copper alloys and give their typical compositions.

Q6: List at least five important non-ferrous metals commonly used in machine design.

Q7: State at least 5 important mechanical properties of materials to be considered in the machine design.

Q8: Define resilience and discuss its implication in the choice of materials in the machine design.

Section Ⅴ Supplementary Reading

Mechanical Properties of Engineering Materials

Choice of materials for a machine element depends very much on its properties, cost, availability and such other factors. It is therefore important to have some idea of the common engineering materials and their properties before learning the details of the design procedure.

The important properties from the design point of view are as follows.

Strength: The strength of a material refers to the material stability to withstand an applied stress without failure. Yield strength refers to the point on the engineering stress-strain curve (as opposed to true stress-strain curve) beyond which the material begins deformation that cannot be reversed upon removal of the loading. Ultimate strength refers to the point on the engineering stress-strain curve corresponding to the maximum stress. The applied stress may be tensile, compressive, or shear.

Hardness: Property of the material that enables it to resist permanent deformation, penetration, indentation, etc. Sizes of indentations by various types of indenters are the measure of hardness, e. g., Brinnel hardness test, Rockwell hardness test, Vickers hardness (diamond pyramid) test. These tests give hardness numbers which are related to yield pressure (MPa).

Elasticity: This is the property of a material to regain its original shape after deformation when the external forces are removed. All materials are plastic to some extent but the degree varies, for example, both mild steel and rubber are elastic materials but steel is more elastic than rubber.

Plasticity: This is associated with the permanent deformation of material when the stress level exceeds the yield point. Under plastic conditions, materials ideally deform without any increase in stress.

Ductility: This is the property of the material that enables it to be drawn out or elongated to an appreciable extent before rupture occurs. The percentage elongation or percentage reduction in area before rupture of a test specimen is the measure of ductility. Normally, if percentage elongation exceeds 15%, the material is ductile and if it is less than 5%, the material is brittle. Lead, copper, aluminium, mild steel are typical ductile materials.

Malleability: It is a special case of ductility where it can be rolled into thin sheets but it is not necessary to be so strong. Lead, soft steel, wrought iron, copper and aluminium are some materials in order of diminishing malleability.

Brittleness: This is opposite to ductility. Brittle materials show little deformation before fracture and failure occur suddenly without any warning. Normally, if the elongation is less than 5%, the material is considered to be brittle, e. g., cast iron, glass, ceramics are typical brittle materials.

Resilience: This is the property of the material that enables it to resist shock and impact by storing energy. The measure of resilience is the strain energy absorbed per unit volume.

Toughness: This is the property which enables a material to be twisted, bent or stretched under impact load or high stress before rupture. It may be considered to be the ability of the material to absorb energy in the plastic zone. The measure of toughness is the amount of energy absorbed after being stressed up to the point of fracture.

Creep: When a member is subjected to a constant load over a long period of time, it undergoes a slow permanent deformation and this is termed as "creep". This is dependent on temperature. Usually, at the elevated temperatures, the creep is high.

Exercises

1. Write T(True) or F(False) beside the following statements about the text.

(1) ________ Common engineering materials are normally classified as the metals. Metals may conveniently be divided into ferrous metals and alloys.

(2) ________ Copper is tough, malleable and ductile and can easily be forged or welded. It cannot however take sudden shocks.

(3) ________ Steel is by far the most important engineering material and there is an enormous variety of the steel to meet a wide variety of engineering requirements.

(4) ________ Chief alloying elements added are usually nickel for strength and toughness, chromium for hardness and strength, tungsten for hardness at elevated temperature, vanadium for tensile strength, manganese for high strength in hot rolled and heat treated condition, silicon for high elastic limit.

(5) ________ Metals containing elements other than steel, carbon and nickel as their chief constituents are usually referred to as non-ferrous metals.

(6) __________They are the materials which can be forged into the desired shapes under the impact with or without the application of the heat.

(7) __________ It is therefore important to have some idea of the common engineering materials and their properties after learning the details of the design procedure.

(8) __________The strength of a material refers to the material resistance to withstand an applied stress when the failure occurs.

2. Choose the best answer.

(1) Metals may conveniently be divided into__________ and__________ metals.

A. conductible/non-conductible B. ferrous/non-ferrous

C. light/heavy D. black/white.

(2) Cast iron is an__________ of iron, carbon and silicon and it is hard and brittle.

A. alloy B. wood C. stone D. plastic

(3) The properties of plain carbon steel depend mainly on the__________ percentages and other alloying elements are not usually present in more than 0.5% to 1% such as 0.5% Si or 1% Mn, etc.

A. wood B. metal C. ferrous hardening D. carbon

(4) Non-metallic materials are also used in the__________ practice due to, principally, their low cost, flexibility and resistance to the heat and electricity.

A. engineering B. aeronautics C. marine D. biochemical

3. Fill in the blanks with words or expressions according to the text.

hardness and strength, brittleness, industrial, property, yield strength, exceed

(1) __________refers to the point on the engineering stress-strain curve (as opposed to true stress-strain curve) beyond which the material begins deformation that cannot be reversed upon removal of the loading.

(2) It is therefore important to have some idea of the common engineering materials and their __________before learning the details of the design procedure.

(3) It provides hardness but tin content also oxidizes resulting in__________.

(4) Normally, if percentage elongation __________ 15% the material is ductile and if it is less than 5% the material is brittle.

(5) Steel is basically an alloy of iron and carbon in which the carbon content can be less than 1.7% and carbon is present in the form of iron carbide to impart __________.

(6) They are now extensively used in various __________ applications for their corrosion resistance, dimensional stability and relatively low cost.

4. Answer the following questions according to the text.

(1) 金属材料是用来做什么的？常见的金属都有哪些？

(2) 合金钢中加碳是为了什么？合金中加入的哪些元素是为了获得良好的韧度和硬度？

(3) 非金属材料为什么可以用于工程实践？塑料为什么能广泛应用？

5. Translate the following paragraphs into Chinese.

(1) Non-metals are also used in the engineering practice due to, principally, their low cost, flexibility and resistance to the heat and electricity. Though there are many suitable non-metals, plastics is the most important from the design point of view. They are synthetic materials which can be moulded into the desired shapes under pressure with or without the application of the heat. They are now extensively used in various industrial applications for their corrosion resistance, dimensional stability and relatively low cost.

(2) Choice of materials for a machine element depends very much on its properties, cost, availability and such other factors. It is therefore important to have some idea of the common engineering materials and their properties before learning the details of the design procedure.

(3) This is the property which enables a material to be twisted, bent or stretched under impact load or high stress before rupture. It may be considered to be the ability of the material to absorb energy in the plastic zone. The measure of toughness is the amount of energy absorbed after being stressed up to the point of fracture.

单元评价

通过本单元的学习，使学生掌握工程材料与热处理相关的英语词汇及常用表达语句，并具备运用英语检索工程材料与热处理相关信息及使用英语描述工程材料与热处理相关内容的能力，同时培养学生吃苦耐劳的工作精神、认真负责的工作态度及团队合作与交流的能力。

单元 3

电子电气应用

【学习目标】

知识目标：

1. 熟悉电子电气应用的相关英语词汇。
2. 掌握与人交流电子电气技术应用时常用的英语表达语句。

技能目标：

1. 具备电子电气应用英语词汇寻读和跟读技巧，能运用英语检索相关信息。
2. 具备使用英语描述电子电气应用相关信息的能力。

素养目标：

1. 培养学生制造强国、科技强国的使命担当。
2. 培养学生团队协作、吃苦耐劳、无私奉献的匠心品质。
3. 培养学生爱岗敬业、精益求精、不断创新的工匠精神。

Section I Texts

Lift control system is used to manipulate each control process by managing such commands as running direction, car call, load signal, landing indication, safety protection.[1] Lifts in different applications have different load, speed and drive/control modes. Lift in the same application may also have different control mode. Whatever control mode is adopted, the objective is the same, that is to say, to be specific, according to car call and landing call, lift control system will execute automatic logic judgment to determine which lift will receive signal and the direction lift will run towards and complete programmed control objective through the electrical automatic system based on the command.[2] Fig. 3-1-1 shows an electrical control box.

1. Types of Lift Control System

Control system development chronicle indicates that there has appeared many control modes, such as relay control, PLC, single chip computer control, multiple-computer control. Prevailing in different era, these control modes are still employed in lifts, now, due to massive integrated circuit and computer technology development.[3]

Fig. 3-1-1 Electrical control box

2. Lift Control Function

Many functions are involved from the time landing call is pressed to the time when passengers get out of the car, such as response to call, start-up, running, stop, door-open/close. In the design of lift control system, some functions are standard, while some are optional, which customers can select. Below is the introduction of typical control functions for collective selective lift.

1) Response to car call

Car call and landing call are referred to as call signals.

(1) Response to car call.

① If car call is on the same direction as car running, car will respond according to the order.

② If car is dwelling on a certain landing or decelerating towards a certain landing, car call of the landing will not be registered.

③ If a car call is registered, car call indicator or LED will light up and display.

④ Car call will be canceled at door zone.

(2) Response to landing call.

① If landing call is on the same direction as car running, car will response according to the order.

② The last reverse call signal will also be responded to.

③ If landing call is in reverse direction to car running, the call will be reserved. After responding to the last call in the same direction, car will respond to landing call in the reverse direction according to the order.

④ If landing call from a certain landing on which car is dwelling is in the same direction with car running, door can re-open; if this landing call is in reverse direction with car running, it will be reserved. The prerequisites are: car arrives at a certain landing and reserves its running direction, or its running direction has been preset.

⑤ Landing call will be canceled under the following circumstances:

If the landing call is registered before the deceleration point, the landing call will be canceled on the deceleration point.

If the landing call is registered after the deceleration point, the landing call will be canceled in the door zone.

⑥ If up/down call buttons are pressed in the middle landing, the call in the same direction as the car running will be canceled as soon as the car arrives at the landing, while the call on the reverse direction will be reserved.

⑦ If the load is full, the full load indicator will light up. Landing call will be reserved rather than being responded.

⑧ After the landing call is registered, the landing call indicator or LED will be lighted up. Fig. 3-1-2 shows response to landing call.

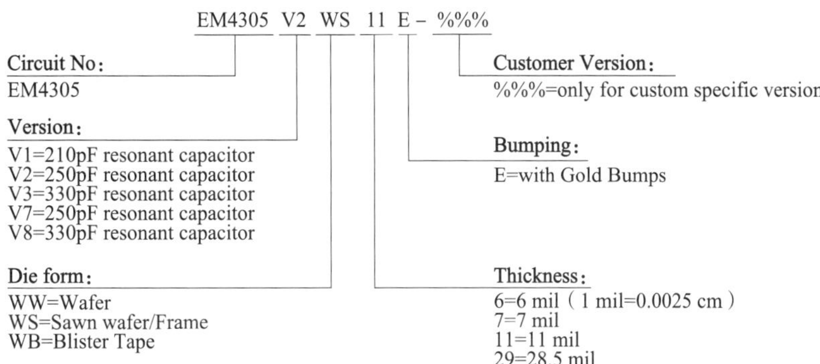

Fig. 3-1-2 Response to landing call

2) Home landing parking

(1) After responding to the last call, the car will return to home landing if no calls are received in a period of time.

(2) After returning to home landing, the car will remain waiting with the car door closed.

(3) In the course of returning to the home landing, the car will respond to call from following landing that is in the same direction as the car running; the car will not respond to the call from preceding landing and park at the nearest landing, with the door closed and will respond to the call by running in the reverse direction. Fig. 3-1-3 shows the base station docking system.

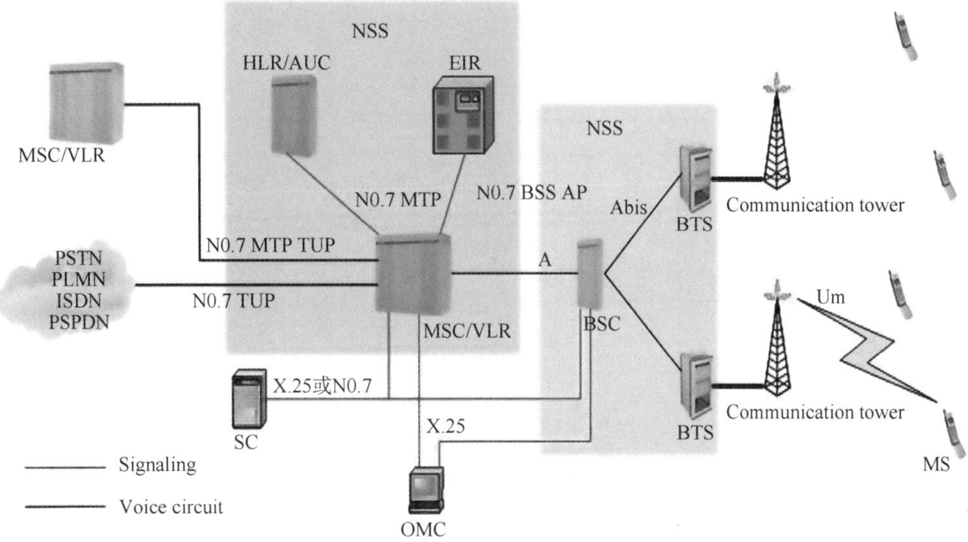

Fig. 3-1-3 Base station docking system

3) Dwelling

(1) When connecting with the dwelling switch, lift enters the dwelling status, the car returns to the dwelling landing.

(2) When the car arrives at the dwelling landing, all calls are canceled rather than being registered.

(3) When the car dwells at the dwelling landing, all position indicators, car lights and the fan will be closed, non-operation light will be lighted up.

(4) During the car inspection, the dwelling function will not be activated.

(5) Independent service has a preference over the dwelling operation, but if the car arrives at the dwelling landing, the car door will close and dwell.

(6) In case of only one car call, the anti-crime operation has preference over the dwelling function, but the car will arrive at the dwelling landing to dwell.

4) Emergency stop

(1) Manual emergency stop.

① Normally mounted on the car top, machine room and pit, the emergency switch is used by inspection staff. As non-automatic resets the switch, it must be manually reset once opened so that the lift can restore normal state. Inspection staff can press this switch under the inspection status, making the lift stop and ensuring the personal safety. Fig. 3-1-4 shows elevator emergency braking system.

② In the course of running, the car will stop immediately and cancel all call signals if the emergency switch is opened.

③ During the emergency stop, the car will not register any calls.

④ During the emergency stop, the door will be closed.

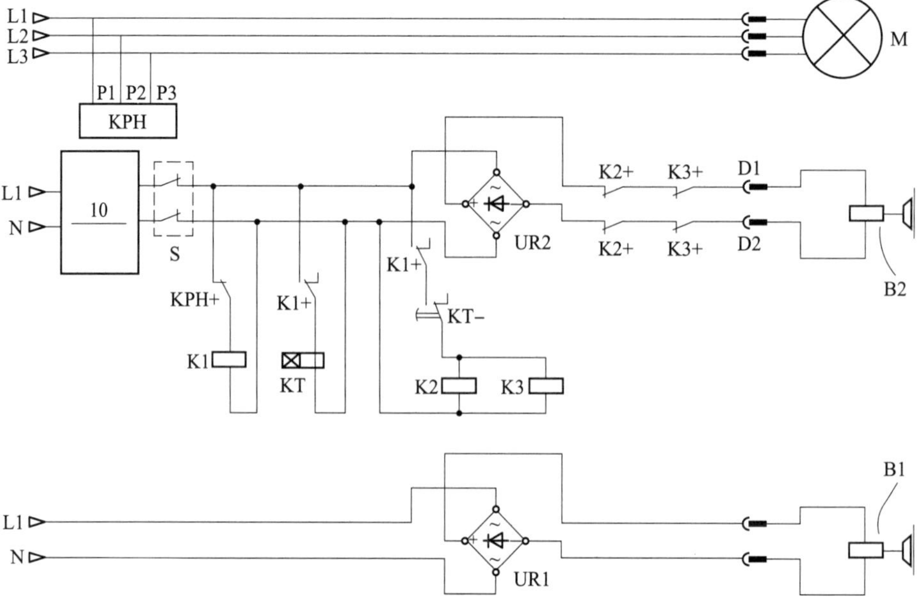

Fig. 3-1-4 Elevator emergency braking system

(2) Failure and emergency stop.

① To ensure the safe operation and protect the passengers against the personal injury, the lift is equipped with a series of safety devices, such as the over speed governor, safety gear, buffer, limit switch, safety window, rope rupture protection and motor overheat protection.[4] Fig. 3-1-5 shows elevator malfunction and emergency stop system.

Should any failure occur, lift emergency stop will be caused by the activation of any safety device and switch or break of contact in the safety circuit.

② After the lift emergency stops, all calls are canceled.

③ After the lift emergency stops, car lighting and fan will not be closed.

5) Independent service

Independent service is mounted in special console of the control panel in a car.

(1) In the course of the running, the car will enter the independent service and stop at the nearest landing if the independent service is connected.

(2) During the independent service, the landing call is not registered.

(3) If car enters independent service during the car stop, the calls are canceled and the door opens.

(4) During the independent service, operate direction button to close the door. When the door is closed, release the button and the door will open again.

(5) During the independent service, the "close" button is ineffective.

(6) During the independent service, the anti-crime is still effective.

6) Anti-crime service

(1) If the anti-crime service is connected, the car will stop at each landing until it arrives at the last landing. Fig. 3-1-6 shows elevator anti-crime service system.

(2) If the anti-crime service switch is disconnected, the anti-crime operation is finished immediately.

(3) During the car running, if the anti-crime service switch is connected, the car will enter anti-crime service operation immediately.

(4) During the lift returning to the home landing, this function is till effective, but after the car arrives at the dwelling landing, this function will be ineffective.

(5) During the independent service, the anti-crime function is still effective.

7) Inspection operation

According the to safety regulations, an accessible control device is mounted on the car top to facilitate the inspection and maintenance, (Fig. 3-1-7) i. e., the inspection box on the car top.[5] For most lifts, an inspection box is mounted in the machine room and on the car top. Inspection switch is a double stable electrical safety switch, which can be in the inspection and normal operation positions, with the abuse protection.

(1) Car top inspection has the preference over the machine room inspection, i. e., if the car top inspection switch is switched to the inspection position, the machine room inspection switch will be ineffective.

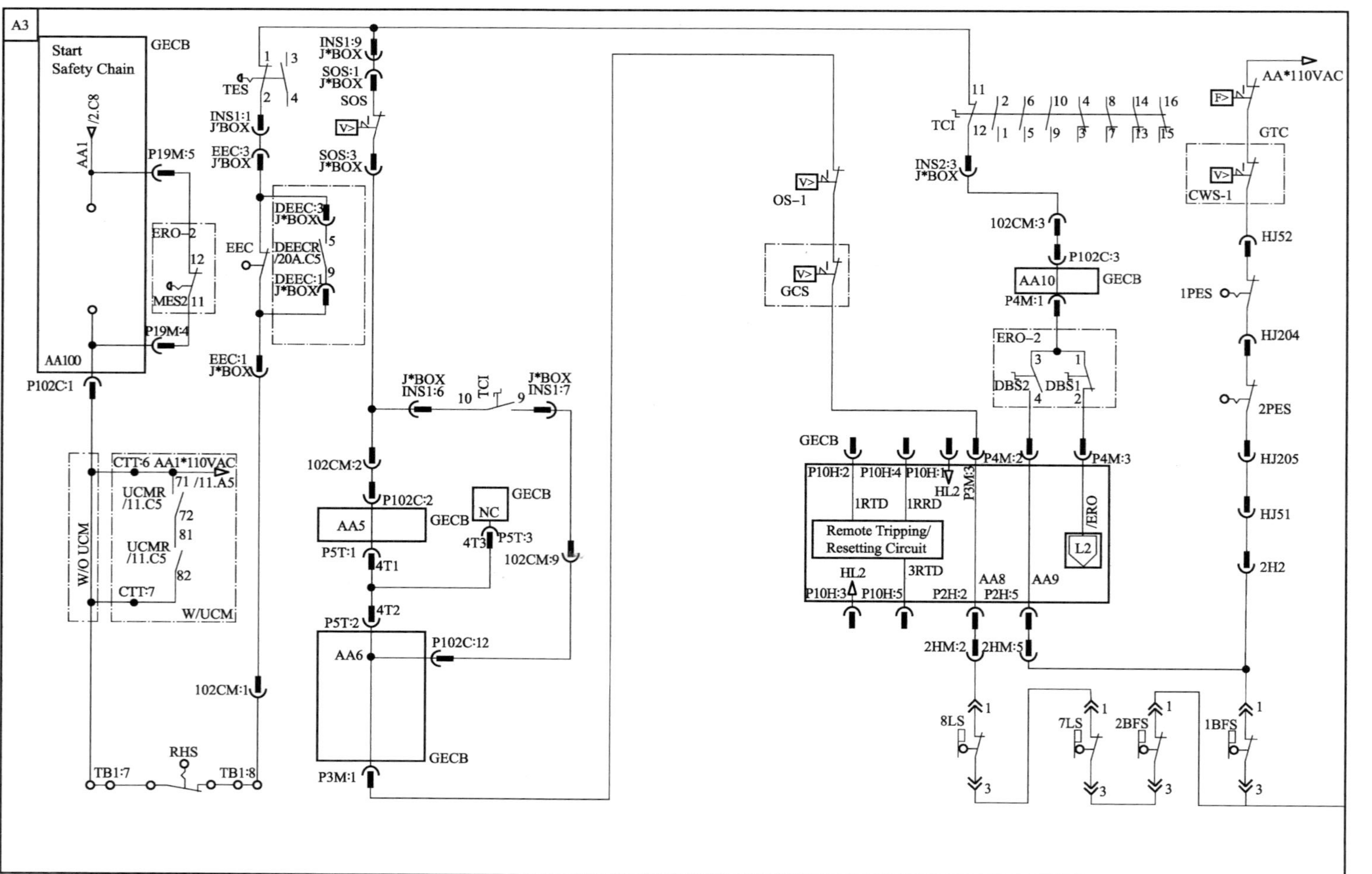

Fig. 3-1-5 Elevator malfunction and emergency stop system

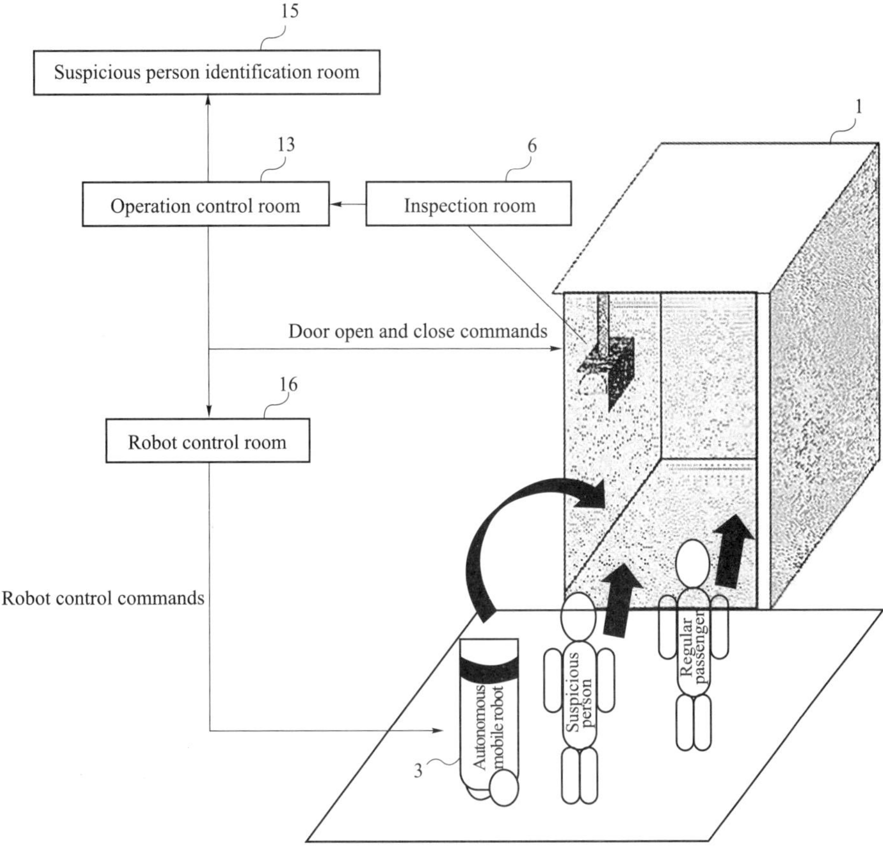

Fig. 3-1-6 Elevator anti-crime service system

Fig. 3-1-7 Maintenance and repair of elevators

(2) Only if the car top inspection switch is in the normal operation switch, the machine room inspection switch will be effective.

(3) Once the inspection switch is switched to the inspection position, the lift will enter the inspection mode, all normal operation will be canceled, including any automatic door operation. Inspection will complete until the inspection switch is switched to the normal operation switch again.

(4) When the car is running, the car emergency stop will occur and enter the inspection mode if the inspection switch is switched to the inspection position.

(5) During the inspection, the lift runs according to the push-in buttons indicating up or down; press the push-in button, the lift will run at the inspection speed.

(6) During the lift inspection operation, all safety devices should be effective.

8) Rescue operation

If the power is cut off and the lift is running outside the leveling zone, the control system will make the lift enter slow leveling status when the power is on again. Consequently, the lift will arrive at the leveling position slowly with the door opens, then resume to the normal status.

Elevator maintenance unit provide mainteinance and emergency rescue (Fig. 3-1-8) services to elevator user unit.

Fig. 3-1-8 Emergency rescue of elevators

9) Releveling

(1) Car will relevel if the lift is in the door zone with the leveling error of rate ±15mm and receives the releveling signal from the control system.

(2) If the car is above the leveling position by 15mm, it will relevel by moving downwards; if the car is below the leveling position by 15mm, it will relevel by moving upwards.

(3) When door fully opens, car will begin releveling; releveling completes after the releveling signal disappears.

10) Alarm

When the failure or emergency occurs on the lift, the passengers can inform the attendants of the situation by pressing the alarm button (Fig. 3-1-9) or other device for rescue.

11) Lighting circuit

Lighting circuit is separately set with the main circuit so that the lighting will be on to facilitate the rescue and maintenance, should main power supply fail. Fig. 3-1-10 shows an elevator lighting circuit.

Fig. 3-1-9 Elevator alarm

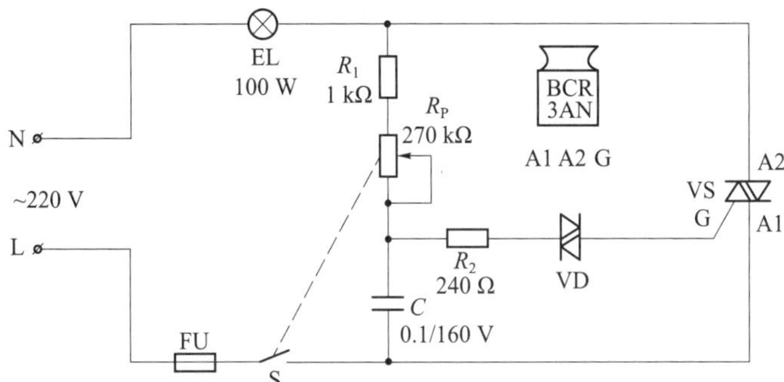

Fig. 3-1-10 Elevator lighting circuit

12) Fireman service

Located near the lift key switch in the home landing, the fireman service switch is normally enclosed by the glass board and not allowed to be pressed. When the building is on fire, the fireman service switch can be pressed after the glass board is crushed in order to make the lift enter the fireman service status.

(1) Fireman service returning to home landing.

① Car calls and landing calls are canceled.

② If the car is running towards the home landing, the car will arrive at home landing nonstop.

③ If the car is running opposite to the home landing, the car will stop at the nearest landing, with the door closed, then run directly to the home landing.

④ If the car is on a certain landing with the door open, it will return to the home landing after closing the door; if the car is on a certain landing with the door closed, it will return to the home landing immediately.

⑤ After the car arrives at the home landing with the door open, the door will close after a preset time delay.

⑥ If the car is on home landing waiting, it will enter the fireman service special status.

(2) Fireman special status.

① Landing call remains canceled;

② Car call button command is restored to facilitate the fireman to operate;

③ During the fireman service, the door open button is still effective.

④ Automatic returning to the home landing function is canceled.

(3) Anti-abuse service.

① When the car load is less than 80 kg and the car calls are less than 4, the car will arrive at a certain landing with the door open and all signals canceled.

② If car calls are more than 4, all the car call signals will be canceled.

③ During the independent service, this function will be ineffective.

(4) Door open/close.

Passenger lifts should have the automatic door open/close functions.

① After the leveling, the door will automatically open. The door will automatically close after a specific period of time delay that can be adjusted.

② Control panel is equipped with the door open/close buttons (Fig. 3-1-11). Manually press these buttons to close or open the door.

③ When a car that will continue to go toward a preset direction is on a landing, the door will open if the landing call button is pressed, while the car is in the same direction as the car running.

④ If the car is on a certain landing waiting, the landing call button can be pressed to open the door.

⑤ In the course of closing, the door will re-open if the safety edge (door vane) or the light curtain (ray protection) touches the obstruction.

⑥ During the door closing, the door open button can be pressed to re-open the door.

⑦ In case of the car overload, the overload indicator will light up and the buzzer will sound; lift will not close the door or the door will re-open in the course of the closing until the overload is eliminated.

Fig. 3-1-11 Opening and closing doors of elevators

Section Ⅱ New Words and Phrases

manipulate *vt.* 操纵，操作

indication *n.* 指示，显示

car call	轿厢内指令信号
landing call	层站召唤信号
relay control	继电器控制
massive integrated circuit	大规模集成电路
PLC	可编程式逻辑控制器
decelerate	*v.* 减速
reverse	*v.* 反转
prerequisite	*n.* 先决条件，前提
dwelling	*n.* 驻停
emergency stop	紧急停车
manual emergency stop	人为紧急停车
overspeed governor	限速器
safety gear	安全钳
rope rupture protection	断绳保护
anti-crime service	防犯罪服务
inspection operation	检修操作
safety regulation	安全规范
accessible	*adj.* 易接近的
double stable electrical safety switch	双稳态开关
cut off	切断
consequently	*adv.* 因此
alarm	*vt.* 报警 *n.* 警报
lighting circuit	照明电路
fireman service	消防服务
fireman special status	消防员专用状态
anti-abuse service	防捣乱服务
safety edge	安全触板
light curtain	光幕
ray protection	光保护装置

Section Ⅲ Notes to Complex Sentences

[1] Lift control system is used to manipulate each control process by managing such commands as the running direction, car call, load signal, landing indication, safety protection.

电梯的控制主要指对电梯在运行过程中的运行方向、层站召唤、负载信号、楼层显示、安全保护等指令信息进行管理，操纵电梯实行每个控制环节的方式和手段。

[2] Whatever control mode is adopted, the objective is the same, to be specific,

according to the car call and landing call, the lift control system will execute the automatic logic judgment to determine which lift will receive the signal, which direction the lift will run towards and complete the programmed control objective through the electrical automatic system based on the command.

不论电梯使用何种控制方式，所要达到的目标是相同的，具体来说就是根据轿厢内的指令信号、层站召唤信号而自动进行逻辑判定，决定哪一台电梯接收信号，自动定出电梯的运行方向，并按照指令要求通过电气自动控制系统完成预定的控制目的。

[3] Control system development chronicle indicates that there has appeared many control modes, such as the relay control, PLC, single chip computer control, multiple-computer control. Prevailing in different era, these control modes are still employed in lifts now due to the massive integrated circuit and the computer technology development.

从控制系统的实现方法来看，电梯的控制系统经历了继电器电梯控制、可编程序控制器控制、单片机控制、多微机控制多种形式，这些控制方式代表了 不同时期电梯控制系统的主流，目前的在用电梯中都有应用，并且随着大规模集成电路和计算机技术的发展而逐步推陈出新。

[4] To ensure the safe operation and protect the passengers against the personal injury, the lift is equipped with a series of safety devices, such as the overspeed governor, safety gear, buffer, limit switch, safety window, rope rupture protection, motor overheat protection.

电梯设有一系列安全保护装置，如限速器、安全钳、缓冲器、极限开关、安全窗、断绳保护、电机过热保护等以保证电梯的安全运行，从而保护乘客的人身安全。

[5] According to the safety regulations, an accessible control device is mounted on the car top to facilitate the inspection and maintenance, i. e. inspection box on the car top.

根据电梯安全规范，为便于检修和维护，应在轿顶装一个易于接收信号的控制装置，即轿顶检修盒。

Section Ⅳ Exercises

Translate the following paragraphs into Chinese.

The Battery Bus

To the everyday bus operator, battery electric power would seem to offer few specific advantages over the conventional diesel power. Specific objections are invariably ones of short range, low performance and higher primary energy consumption. Advantages are often mooted ones of low emission and therefore low pollution in urban environs, low noise pollution, greater passenger comfort and consumption of “off peak” surplus the electrical energy.

The ideal battery for the electric buses should be able to offer:

(1) high energy density;

(2) high power density;

(3) high electrochemical efficiency;

(4) long service life and low maintenance requirements;

(5) economical energy costs;

(6) safety to operate and serve;

(7) no temperature control problems in the battery (e. g., overheating) or high running temperature.

Unfortunately, no battery system presently in the commercial production can offer all of these facilities.

Section V Supplementary Reading

What is the ideal battery for the electric buses like?

Clearly the electric bus is, in the short term, a practical proposition only in the "in-city" role; longer rural or inter-city operation is out of the question now and is likely to be for a very long time to come.

One specific barrier to the wider adoption of the electric bus must be the external demand of operators for the "unlimited" mileage—that is, for the vehicles (i. e. diesel engines) with total flexibility. However, there is little to support the view that, in a highly efficient transit operation, almost every vehicle should be capable of running almost every service. The real truth may be that due to the inflexibility of infrastructure, coupled to the low reliability of present technology of diesel buses in almost all the existing high-density public transport operations, a vehicle of high route availability is needed to simply maintain day-to-day service requirements at an adequate level. In this light, maybe the time has come to take a careful look at the "real" economics of that highly complicated engineering package—the diesel bus. Simplicity usually breeds reliability and viewed in this light electric buses may begin to make the sense in the operational, financial and engineering terms.

The Electric Vehicle Association of Great Britain carried out a comparison between electric and diesel vehicles in the already well-established UK applications of the electric vehicles, primarily the milk floats and light vans. From this, the conditions that make the electric vehicles viable are:

(1) the long life of the electric vehicle; fifteen years against five for the conventional vehicle, which goes to offset the higher capital cost of the electric over the conventional vehicle;

(2) the tax advantages, i. e., no road tax and no tax on fuel;

(3) lower maintenance requirement.

One aspect of the bus financing could operate in favor of the electric bus (Fig. 3-5-1). Battery bus purchasers at present receive a capital grant from the government. Given a higher capital cost for the electric buses, the cash grant payable would be higher for a conventional vehicle. However, this bus grant is rapidly being phased out.

Fig. 3-5-1 Electric bus

Exercises

1. Write T(True) or F(False) beside the following statements about the text.

(1) ________ Lift control system is used for the process by managing such commands as turning the direction, emergency call, pressure signal, departure indication, firefighting, etc.

(2) ________ Control system development chronicle indicates that there has appeared many control modes, such as relay control, PLC, single chip computer control, multiple-computer control.

(3) ________ In the design of lift control system, some functions are standard, while some are not useful, which customers can give up.

(4) ________ If landing call from a certain landing on which car is dwelling is on the same direction as car running, door can re-open; if this landing call is in reverse direction with car running, it will be reserved.

(5) ________ When disconnecting with the dwelling switch, lift goes out of the dwelling status, the car returns to the dwelling landing.

(6) ________ One specific barrier to the wider adoption of the electric switch must be the internal demand of operators for the "unlimited" mileage—that is, for the vehicles (i. e., diesel engines) with the single flexibility.

(7) ________ The long life of the electric vehicle; fifteen years against five for the conventional vehicle, which goes to offset the higher mechanical and labor cost over the conventional vehicle.

(8) ________ Given a higher capital cost for the electric buses, the cash grant payable would be higher for a conventional vehicle. However, this bus grant is rapidly being phased out.

2. Choose the best answer.

(1) The lift control system will execute automatic logic judgment to determine which lift will receive signal and the direction lift will run towards and complete programmed control objective through the ________ based on the command.

A. manual control system　　B. intelligent control system

C. fully automatic system　　D. electric automated system

(2) If landing call is on the same direction as car running, car will response according to ________.

A. demand B. requirement C. instruction D. order

(3) In the course of________, the door will re-open if the safety edge (door vane) or the light curtain (ray protection) touches the obstruction.

A. open B. stop C. closing D. station

(4) Specific objections are invariably ones of short range, low performance and higher primary________.

A. energy charge B. energy conversion C. energy storing D. energy consumption

(5) In this light, maybe the time has come to take a careful look at the "real" economics of that highly complicated engineering package—the________bus.

A. new energy B. new type C. petrol D. diesel

3. Fill in the blanks with words or expressions accrding to the text.

(1) According the to________, an accessible control device is mounted on the car top to facilitate the ________, i. e., the inspection box on the car top.

(2) Only if the car top inspection switch is in the normal________ switch, the machine room inspection switch will be________.

(3) If the power is cut off and the lift is running outside the________, the control system will make the lift enter slow leveling status when the power is on again, consequently, the lift will arrive at the leveling position slowly with the door open, then resume to the________.

(4) ________ is separately set with the main circuit so that the lighting will be on to facilitate the________, should main power supply fail.

(5) When a car that will continue to go toward________is on a landing, the door will open if ________is pressed, while the car is on the same direction as the car running.

(6) The long life of the________; fifteen years against five for the conventional vehicle, which goes to offset the higher capital cost of the electric over the ________.

4. Answer the following questions according to the text.

(1) 电梯是怎么控制的？电梯的内选信号和外呼信号是如何应答的？

(2) 电梯的检修是如何操作的？电梯有哪些故障与急停保障设施？

(3) 电梯是如何开关门的？商用电梯和小区居民楼的电梯是否一样，为什么？

5. Translate the following paragraphs into Chinese.

(1) Control system development chronicle indicates that there has appeared many control modes, such as relay control, PLC, single chip computer control, multiple-computer control. Prevailing in different era, these control modes are still employed in lifts, now, due to massive integrated circuit and computer technology development.

(2) If landing call from a certain landing on which car is dwelling is on the same direction as car running, door can re-open; if this landing call is on reverse direction with car running, it will be

reserved. The prerequisites are: car arrives at a certain landing and reserves its running direction, or its running direction has been preset.

单元评价

通过本单元的学习，使学生掌握了电子电气应用相关的英语词汇及常用表达语句，并具备运用英语检索电子电气应用相关信息及使用英语描述电子电气应用相关内容的能力，同时培养学生爱岗敬业、精益求精、不断创新的工匠精神。

单元 4

机械原理与机械设计

【学习目标】

知识目标：

1. 熟悉机械原理与机械设计的相关英语词汇。
2. 掌握与人交流机械原理与机械设计时常用的英语表达语句。

技能目标：

1. 具备机械原理与机械设计英语词汇寻读和跟读技巧，能够运用英语检索相关信息。
2. 具备使用英语描述机械原理与机械设计相关信息的能力。

素养目标：

1. 培养学生创新意识和法治意识。
2. 培养学生钻研精神和精益求精的工匠精神。
3. 提高学生团队协作和解决问题的能力。

Section Ⅰ Texts

Design is essentially a decision-making process. If we have a problem, we need to design a solution. In other words, to design is to formulate a plan to satisfy a particular need and to create something with a physical reality. A bad decision leads to a bad design and a bad product.

There are many factors to be considered while attacking a design problem. In many cases, this is a common sense approach to solving a problem.[1] Some of these factors are as follows.

(1) Device or mechanism to be used. This is best judged by understanding the problem thoroughly. Sometimes, a particular function can be achieved by a number of means or by using different mechanisms and the designer has to decide which one is the most effective under the circumstances.

(2) Material. This is a very important aspect of any design. A wrong choice of material may lead to failure, over or undersized product or expensive items.[2] The choice of materials is thus dependent on suitable properties of the material for each component, their suitability of fabrication

or manufacture and the cost.

(3) Load. The external loads cause internal stresses in the elements and these stresses must be determined accurately since these will be used in determining the component size.

(4) Size, shape, space requirements and mass—Preliminary analysis would give an approximate size. But if a standard element is to be chosen, the next larger size must be taken.[3] Shapes of standard elements are known but for non-standard element, shapes and space requirements must depend on available space in a particular machine assembly. A scale layout drawing is often useful to arrive at an initial shape and size.[4] Mass is important depending on application.

(5) Manufacture. Care must always be taken to ensure that the designed elements may be manufactured with ease, within the available facilities and at low cost.

(6) How will it operate. In the final stage of the design, a designer must ensure that the machine may be operated with ease. In many power operated machines, it is simply a matter of pressing a knob or switch to start the machine.[5] However, in many other cases, a sequence of operations is to be specified. This sequence must not be complicated and the operations should not require excessive force.

(7) Reliability and safety. Reliability is an important factor in any design. A designed machine should work effectively and reliably. The probability that an element or a machine will not fail in use is called reliability. Reliability lies between $0<R<1$. To ensure this, every detail should be examined. Possible overloading, wear of elements, excessive heat generation and other such detrimental factors must be avoided. There is no single answer for this but an overall safe design approach and care at every stage of design would result in a reliable machine.

Safety has become a matter of paramount importance these days in design. Machines must be designed to serve mankind, not to harm it. Industrial regulations ensure that the manufacturer is liable for any damage or harm arising out of a defective product.[6]

(8) Maintenance, cost and aesthetics—Maintenance and safety are often interlinked. Good maintenance ensures good running condition of machinery. Often a regular maintenance schedule is maintained and a thorough check up of moving and loaded parts is carried out to avoid catastrophic failures. Low friction and wear is maintained by proper lubrication. This is a major aspect of design since wherever there are moving parts, friction and wear are inevitable. High friction leads to increased loss of energy. Wear of machine parts leads to loss of material and premature failure.

Cost and aesthetics are essential considerations for product design. Cost is essentially related to the choice of materials which in turn depends on the stresses developed in a given condition.[7] Although in many cases aesthetic considerations are not essential aspects of machine design, ergonomic aspects must be taken into considerations. Fig. 4-1-1 shows conventional mechanical principles and clesign.

Fig. 4-1-1 Conventional mechanical principles and design

Section Ⅱ New Words and Phrases

decision-making	决策
formulate	*vt*. 制订
thoroughly	*adj*. 彻底地
circumstance	*n*. 情况
undersized	*adj*. 欠尺寸的，尺寸不够的
component	*n*. 组件，元件
external	*adj*. 外部的
internal	*adj*. 内部的
stress	*n*. 应力
approximate	*adj*. 近似的
standard element	标准件
scale	*n*. 比例
scale layout drawing	按比例的设计图
common sense approach	常识性的方法
with ease	轻而易举
facility	*n*. 设施
knob	*n*. 把手，旋钮
sequence	*n*. 顺序，序列
excessive	*adj*. 过分的，过度的
reliability	*n*. 可靠性
probability	*n*. 可能性，概率

overload	*vt.* 过载
excessive heat	过热
preliminary	*adj.* 初步的
assembly	*n.* 装配
paramount	*adj.* 至高无上的
detrimental	*adj.* 不利的，有害的
industrial regulation	工业法规
liable	*adj.* 对……负责的
defective	*adj.* 有缺陷的
aesthetics	*n.* 美学，审美
maintenance	*n.* 维护，维修
catastrophic	*adj.* 灾难性的
friction and wear	摩擦与磨损
lubrication	*n.* 润滑
check up	检查
inevitable	*adj.* 不可避免的
premature	*adj.* 过早的
ergonomics	*n.* 人体工程学

Section Ⅲ Notes to Complex Sentences

[1] In many cases, this is a common sense approach to solving a problem.

许多场合下，通过常识就可解决问题。

这里，approach to solving a problem 意为“解决问题的途径”；common sense 在此处作定语，修饰 approach.

[2] A wrong choice of the material may lead to the failure, over or undersized product or expensive items.

选材错误会导致产品失败、尺寸过大或过小及成本过高。

这里，over or…与前面的 failure 并列，作为 lead to 的宾语。

[3] But if a standard element is to be chosen, the next larger size must be taken.

如果选用标准件，就要取稍大一级的尺寸。

这里，the next larger size 指稍大一级的标准件尺寸。

[4] A scale layout drawing is often useful to arrive at an initial shape and size.

按比例画出一张设计图往往对形成初步的形状和尺寸很有帮助。

句中，scale 意为“比例”；arrive at 意为“形成，达到”。

[5] …it is simply a matter of pressing a knob or switch to start the machine.

……只不过是按一下旋钮开机那么简单的事情。

句中，it is a matter of...意为“是……的事情”。

[6] Industrial regulations ensure that the manufacturer is liable for any damage or harm arising out of a defective product.

工业法规确保制造商对有缺陷产品造成的任何损害和伤害负责。

句中，arising out of...意为“由……产生的”。

[7] Cost is essentially related to the choice of materials which in turn depends on the stresses developed in a given condition.

成本最终取决于材料的选择，而材料的选择又取决于给定条件下的应力状况。

句中，which 引出定语从句，作主语，等同于前面的 choice of materials; in turn 意为“依次”，此处可译为“又”。

Section Ⅳ Exercises

Translate the following paragraphs into Chinese.

Machine design may be classified by the following 3 types.

(1) Adaptive design.

This is based on the existing design, for example, the standard products or systems adopted for a new application. Conveyor belts, control system of the machines and mechanisms or the haulage systems are some of the examples where existing design systems are adapted for a particular use.

(2) Developmental design.

Here we start with an existing design but finally a modified design is obtained. A new model of a car is a typical example of a developmental design.

(3) New design.

This type of design is an entirely new one but based on existing scientific principles. No scientific invention is involved but requires the creative thinking to solve a problem. Examples of this type of design may include designing a small vehicle for the transportation of men and material on board a ship or in a desert. Some research activities may be necessary.

Section Ⅴ Supplementary Reading

Factors in Design

The first step in product developments is design. Design is motivated by the following factors.

(1) The need for a specific product.

(2) The level of the available manufacturing technology.

(3) Mass production capability and production cost.

1. Need

Most products are designed and produced because a need has developed for them. With the tremendous capability of the modern industrial manufacturing, needs for the products are easily developed. Therefore, the product designers are engaged in a constant process of designing and redesigning products to fulfill a continuing need.

2. Capability of Available Manufacturing Technology

The capability of the existing manufacturing technology is an extremely important consideration in the product design and development. As the technology develops new and better processes and materials, the designer naturally makes the use of these to design and redesign the products accordingly.

3. Production Capability and Costs

Most products that are made for the general public are only successful if they can be mass produced and marketed at affordable prices. The best product design is only as good as its ability to be competitively produced. Therefore, a designer must consider the following important questions when a product is being developed.

Is the production capability presently available?

If not, will sales of the product justify development of new production technology?

Can the product be manufactured and marketed at a cost that will return investment and profits?

4. Safety and Reliability

Safety and reliability are the critical factors in the design of many products. In recent years, the product safety considerations have attained the greater importance. This has been of the increased public awareness. The designer's responsibility is equal where personal and public safety are concerned.

5. Marketability

If the manufacturing capability exists or can be developed, the next stop involves the marketing and distribution. It has often been started that the invention is one thing, but the distribution is everything. Costs incurred in the marketing include things such as advertising, national or international sales organizations, packaging, distributors and agents, distribution costs, and warranty services. The selling price to the end customer must reflect any or all of these expenses while being low enough to support sales and meet competition.

Exercises

1. Write T(True) or F(False) beside the following statements about the text.

(1) __________ In other words, to design is to formulate a plan to satisfy a particular need and to create something with a physical reality.

(2) ____________ What device or mechanism to be used — This is best judged by understanding the problem thoroughly.

(3) __________ Size, shape, space requirements and mass — Preliminary analysis would give

an exact and precise size.

(4) ________ In many power operated machines, it is simply a matter of switching on switching off the machine.

(5) ________ Possible overloading, wear of elements, excessive heat generation and other such detrimental factors must be strengthened and paid much attention to.

(6) ________ Industrial regulations ensure that the manufacturer is liable for any damage or harm arising out of a good-quality product.

(7) ________ Often a regular maintenance schedule is maintained and a thorough check up of moving and loaded parts is carried out to avoid catastrophic failures.

(8) ________ Although in many cases aesthetic considerations are essential aspects of machine production, ergonomic aspects should not be taken into considerations.

2. Choose the best answer.

(1) A wrong choice of the material may lead to failure, over or ________ product or expensive items.

A. overestimated B. underestimated C. oversized D. undersized

(2) The choice of materials is thus dependent on ________ properties of the material for each component, their suitability of fabrication or manufacture and the cost.

A. adaptable B. suitable C. intended D. unpredictable

(3) In the final stage of the design, a designer must ensure that the machine may be operated with ease. In many ________ machines, it is simply a matter of pressing a knob or switch to start the machine.

A. hand-pushed B. manually operated

C. engine-worked D. power operated

(4) Often a regular ________ schedule is maintained and a thorough check up of moving and loaded parts is carried out to avoid catastrophic failures.

A. repair B. operation C. driving D. maintenance

(5) As the technology develops new and better ________, the designer naturally makes the use of these to design and redesign the products accordingly.

A. process and material B. quality and process

C. process and product D. quality and material

3. Fill in the blanks with words or expressions according to the text.

(1) Sometimes a particular function can be achieved by a number of means or by using different ________ and the designer has to decide which one is the most effective under the ________.

(2) Shapes of standard elements are known but for non-standard element, ________ requirements must depend on available space in a particular ________.

(3) This is a major aspect of ________ since wherever there are moving parts, ________ are inevitable.

(4) The capability of the__________ is an extremely important consideration in __________.

(5) __________ are the critical factors in the design of many products. In recent years__________ have attained the greater importance.

(6) __________incurred in the__________ include such things as advertising, national or international sales organizations, packaging, distributors and agents, distribution costs, and warranty services.

4. Answer the following questions according to the text.

(1) 为什么设计是一个决策过程？实施设计任务时要考虑哪些因素？

(2) 机械设计要考虑机械的安全性和保障性，什么是安全性和保障性？如何确保设计的机械有安全保障？

(3) 影响机械设计最本质的因素是什么？在机械设计中，是否要考虑人体工程学因素？为什么？

5. Translate the following paragraphs into Chinese.

(1) Material — This is a very important aspect of any design. A wrong choice of material may lead to failure, over or undersized product or expensive items. The choice of materials is thus dependent on suitable properties of the material for each component, their suitability of fabrication or manufacture and the cost.

(2) Safety has become a matter of paramount importance these days in design. Machines must be designed to serve mankind, not to harm it. Industrial regulations ensure that the manufacturer is liable for any damage or harm arising out of a defective product.

(3) Adaptive design — This is based on the existing design, for example, the standard products or systems adopted for a new application. Conveyor belts, control system of the machines and mechanisms or the haulage systems are some of the examples where existing design systems are adapted for a particular use.

单元评价

通过本单元的学习，使学生掌握了机械原理与机械设计相关的英语词汇及常用表达语句，并具备了运用英语检索机械原理与机械设计相关信息及使用英语描述机械原理与机械设计相关内容的能力，同时培养学生的创新意识、法治意识及精益求精的工匠精神。

单元 5

零件制作与普通加工

【学习目标】

知识目标：

1. 熟悉零件制作与普通加工的相关英语词汇。
2. 掌握与人交流零件制作与普通加工时常用的英语表达语句。

技能目标：

1. 具备零件制作与普通加工英语词汇寻读和跟读技巧，能够运用英语检索相关信息。
2. 具备使用英语描述零件制作与普通加工相关信息和过程的能力。

素养目标：

1. 培养学生换位思考的方式，养成学生严谨求实的学习态度。
2. 培养学生实事求是、统筹兼顾、适当超前的系统思维方式。
3. 培养学生的探索能力及敬业、专注的工匠精神。

Section Ⅰ Texts

Text 1

1. Gears

Gears(Fig. 5-1-1) are direct contact bodies, operating in pairs, that transmit motion and force from one rotating shaft to another, or from a shaft to a slide (rack), by means of the successively engaging projections called teeth.[1]

Fig. 5-1-1 Gears

(1) Tooth profiles.

Contacting surfaces of the gear teeth (Fig. 5-1-2) must be aligned in such a way that the drive is positive, i. e., the load transmitted must not depend on frictional contact. As shown in the treatment of direct contact bodies, this requires that the common normal to the surfaces not pass through the pivotal axis of either the driver or the follower.

As direct contact bodies, the cycloidal and involute profiles provide both a positive drive and a uniform velocity ratio, i. e., conjugate action.

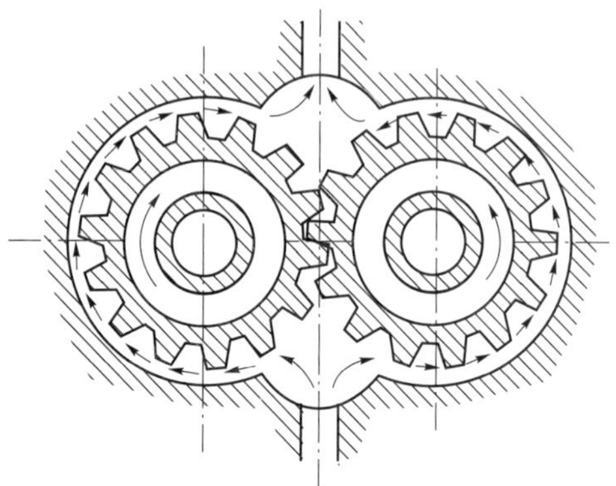

Fig. 5-1-2 Gear teeth meshing

(2) Basic relations.

The smaller of a gear pair is called the pinion and the larger is the gear. When the pinion is on the driving shaft, the pair acts as a speed reducer; when the gear drives, the pair is a speed increaser. Gears are more frequently used to reduce the speed than to increase it.

If a gear having N teeth rotates at n r/min, the product $N\times n$ has the dimension "teeth per minute". This product must be the same for both members of a mating pair if each tooth acquires a partner from the mating gear as it passes through the region of tooth engagement.

For conjugate gears of all types, the gear ratio and the speed ratio are both given by the ratio of the number of teeth on the gear to the number of teeth on the pinion. If a gear has 100 teeth and a mating pinion has 20, the ratio is $100/20=5$. Thus, the pinion rotates five times as fast as the gear, regardless of the speed of the gear. Their point of tangency is called the pitch point, and since it lies on the line of centers, it is the only point at which the tooth profiles have pure rolling contact. Gears on nonparallel, non-intersecting shafts also have pitch circles, but the rolling-pitch-circle concept is not exist.

Gear types are determined largely by the disposition of the shafts. On the one hand, certain types are better suited than others for large speed changes. This means that if a specific disposition of the shafts is required, the type of gear will more or less be fixed. On the other hand, if a required speed change demands a certain type, the shaft positions will also be fixed.

(3) Spur gears and helical gears.

Fig. 5-1-3 show the spur gears and helical gears.

Fig. 5-1-3 Spur gears and helical gears

(a) Spur gear; (b) Helical gear

A gear having tooth elements that are straight and parallel to its axis is known as a spur gear. A spur pair can be used to connect parallel shafts only.

If an involute spur pinion is made of rubber and twists uniformly so that the ends rotates about the axis which is relative to one another, the elements of the teeth which are initially straight and parallel to the axis would become helices.[2] The pinion then would become a helical gear in effect.

(4) Worm and bevel gears.

In order to achieve line contact and improve the load carrying capacity of the crossed axis helical gears, the gear can be made to curve partially around the pinion, in somewhat the same way that a nut envelops a screw. The result would be a cylindrical worm and gear. Instead of being cylindrical, worms are also made in the shape of an hourglass, so that they partially envelop the gear. This results in a further increase in load-carrying capacity.

Worm gears provide the simplest means of obtaining large ratios in a single pair. However, they are usually less efficient than parallel-shaft gears because of an additional sliding movement along the teeth.

2. V-belt

Fig. 5-1-4 shows a V-belt. The rayon and rubber V-belt are widely used for power transmission. Such belts are made in two series, the standard V-belt and the high capacity V-belt. The belts can be used with short center distances and are made endless so that difficulty with splicing devices is avoided.

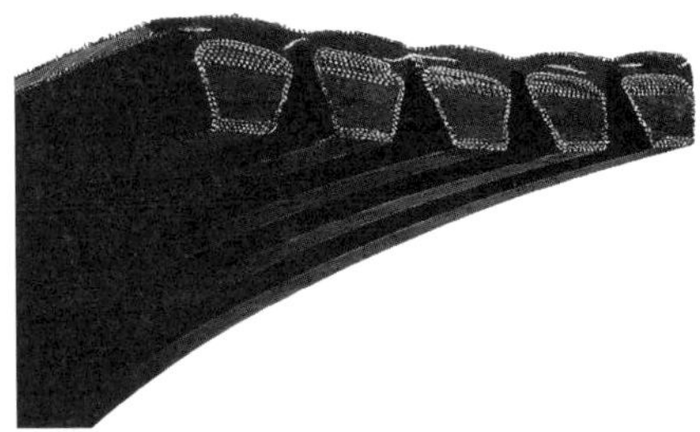

Fig. 5-1-4 V-belt

First, the cost is low and the power output may be increased by operating several belts side by side. All belts in the drive should stretch at the same rate in order to keep the load equally divided among them. When one of the belts breaks, the group must usually be replaced. The drive may be inclined at any angle with tight side either top or bottom. Since belts can operate on relatively small pulleys, large reductions of speed in a single drive are possible.

Second, the included angle for the belt groove is usually 34°–38°. The wedging action of the belt in the groove gives a large increase in the tractive force developed by the belt.

Third, pulley may be made of cast iron, sheet steel, or die-cast metal. Sufficient clearance must be provided at the bottom of the groove to prevent the belt from bottoming as it becomes narrower from wear.[3] Sometimes, the larger pulley is not grooved when it is possible to develop the required tractive force by running on the inner surface of the belt. The cost of cutting the grooves is thereby eliminated. Pulleys are on the market that permit an adjustment in the width of the groove. The effective pitch diameter of the pulley is thus varied, and moderate changes in the speed ratio can be secured.

Exercises

1. Write T(True) or F(False) beside the following statements about the text.

(1) __________ Gears are direct contact bodies, operating in pairs, that transmit motion and force from one rotating shaft to another, or from a shaft to a slide (rack), by means of the successively engaging projections called teeth.

(2) __________ As shown in the treatment of direct contact bodies, this does not require that the common normal to the surfaces should pass through the pivotal axis of both the driver and the follower.

(3) __________ When the pinion is on the driving shaft, the pair acts as a speed reducer; when the gear drives, the pair is a speed increaser.

(4) __________ This product must be the same for either members of a mating pair when each tooth acquires a partner of the mating gear if it passes through the region of teeth engagement.

(5) __________ For conjugate gears of several types, the gear ratio and the speed ratio are not given by the ratio of the number of teeth on the gear or the number of the teeth by the pinion.

(6) __________ Gear types are determined largely by the disposition of the shafts. On the one hand, certain types are better suited than others for large speed changes.

(7) __________ In order to achieve line contact and improve the load carrying capacity of the crossed axis helical gears, the gear can be made to curve partially around the pinion, in somewhat the same way that a nut envelops a screw.

(8) __________ The belts can't be used with short center distances and are not made endless so that difficulty with splicing devices is unavoidable.

2. Choose the best answer.

(1) This product must be the same for both members of a mating pair if each tooth acquires a partner from the ________ as it passes through the region of tooth engagement.

A. mating partner　　B. mating valve　　C. mating chain　　D. mating gear

(2) Their point of tangency is called the________, and since it lies on the line of centers, it is the only point at which the tooth profiles have pure rolling contact.

A. pitch ground　　B. pitch place　　C. pitch point　　D. pitch pot

(3) On the other hand, if a required speed change demands a certain type, the ________ positions will also be fixed.

A. sensor　　B. interface　　C. shaft　　D. flow meter

(4) The result would be a cylindrical worm and gear. Instead of being ________, worms are also made in the shape of an hourglass, so that they partially envelop the gear.

A. cylindrical　　B. hollow　　C. spindle　　D. controlled

(5) All belts in the drive should stretch at the same rate in order to keep the ________ equally divided among them.

A. pressure　　B. load　　C. movement　　D. rotation

3. Fill in the blanks with words or expressions according to the text.

(1) As shown in the treatment of direct contact bodies, this requires that the common normal to the surfaces not pass through________of either the ________or the follower.

(2) When the pinion is on the driving shaft, the pair acts as a ________; when the gear drives, the pair is a ________.

(3) For________of all types, the gear ratio and the ________are both given by the ratio of the number of teeth on the gear to the number of teeth on the pinion.

(4) If ________ is made of rubber and twisted uniformly so that the ends rotates about the axis which is relative to one another, the elements of the teeth which are initially straight and parallel to the axis, would become________.

(5) The belts can be used with short________and are made endless so that difficulty with ________is avoided.

(6) The effective________ of the________ is thus varied, and moderate changes in the speed ratio can be secured.

4. Answer the following questions according to the text.

(1) 齿轮的种类有哪些？齿轮的工作原理是什么？

(2) 大小齿轮的转速关系是什么？

(3) 如何提高V形带的传动功率？

5. Translate the following paragraphs into Chinese.

(1) Gears are direct contact bodies, operating in pairs, that transmit motion and force from

one rotating shaft to another, or from a shaft to a slide (rack), by means of the successively engaging projections called teeth.

(2) If a gear having N teeth rotates at n r/min, the product $N \times n$ has the dimension "teeth per minute". This product must be the same for both members of a mating pair if each tooth acquires a partner from the mating gear as it passes through the region of tooth engagement.

(3) The rayon and rubber V-belt are widely used for power transmission. Such belts are made in two series, the standard V-belt and the high capacity V-belt. The belts can be used with short center distances and are made endless so that difficulty with splicing devices is avoided.

Text 2

Lathes are generally considered to be the oldest machine tools. Although woodworking lathes were originally developed during the period 1000 B. C. —1 B. C., metalworking lathes with lead screws were not built until the late 1700 s. The most common lathe was originally called an engine lathe because it was powered with overhead pulleys and belts from nearby engines. Today these lathes are equipped with individual electric motors. Fig. 5-1-5 shows the ordinary lathe.

Although simple and versatile, an engine lathe requires a skilled machinist because all controls are manipulated by hand. Consequently, it is inefficient for repetitive operations and for large production runs.

Fig. 5-1-5 Ordinary lathe

1. Lathe Components

Lathes are equipped with a variety of components and accessories. The basic components of a common lathe are described below.

(1) Bed.

The bed supports all major components of the lathe. Beds have a large mass and are rigidly built, usually from gray or nodular cast iron. The top portion of the bed has two ways, with various cross-sections that are hardened and machined for wear resistance and dimensional accuracy during use.

(2) Carriage.

The carriage or carriage assembly, slides along the ways and consists of an assembly to the cross-slide, tool post, and apron. The cutting tool is mounted on the tool post (Fig. 5-1-6), usually, with a compound rest that swivels for tool positioning and adjustment. The cross-slide moves radially in and out, controlling the radial position of the cutting tool in operations such as facing.[4] The apron is equipped with mechanisms for both manual and mechanized movement of the carriage and the cross-slide by means of the lead screw.

Fig. 5-1-6 Lathe tool post

(3) Headstock.

The headstock(Fig. 5-1-7) is fixed to the bed and is equipped with motors, pulleys, and V-belts that supply power to the spindle at various rotational speeds. The speeds can be set through manually-controlled selectors. Most headstocks are equipped with a set of gears, and some have various drives to provide a continuously variable speed range to the spindle. Headstocks have a hollow spindle to which workholding devices, such as chucks and collets, are attached, and long bars or tubing can be fed through for various turning operations.

Fig. 5-1-7 Headstock

(4) Tailstock.

The tailstock(Fig. 5-1-8), which can slide along the ways and be clamped at any position, supports the other end of the workpiece. It is equipped with a center that may be fixed (dead center) or may be free to rotate with the workpiece (live center). Drills and reamers can be mounted on the tailstock quill (a hollow cylindrical part with a tapered hole) to drill axial holes in the workpiece.

Fig. 5-1-8 Tailstock

(5) Feed rod and lead screw.

The feed rod is powered by a set of gears from the headstock. It rotates during the operation of the lathe and provides movement to the carriage and the cross-slide by means of gears, a friction clutch, and a keyway along the length of the rod.[5] Closing a split nut around the lead screw engages it with the carriage and it is also used for cutting threads accurately.

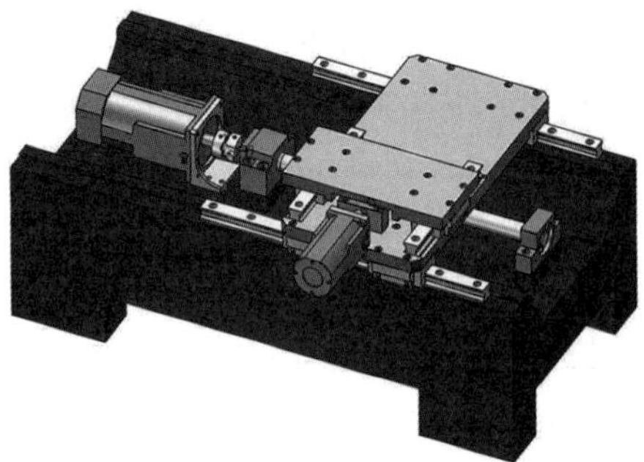

Fig. 5-1-9 The feed system of the lathe

2. Lathe Specifications

A lathe is usually specified by ① its swing, that is, the maximum diameter of the workpiece that can be machined, ② the maximum distance between the headstock and tailstock centers, and ③ the length of the bed. For example, a lathe may have the following size: 360 mm (14 in) swing by 760 mm (30 in) between centers by 1, 830 mm (6ft) length of bed. Lathes are available in a variety of styles and types of construction and power.

Bench lathes are placed on a workbench. They have low power and are usually operated by hand feed, which are used for precision-machine small workpieces.[6] Toolroom lathes have high precision, enabling the machining of parts to close tolerances. Engine lathes are available in a wide range of sizes and are used for a variety of turning operations. In gap bed lathes, a section of the bed in front of the headstock can be removed to accommodate larger-diameter workpieces.

With workpiece sizes as large as 1.7 m in diameter, 8 m in length (66 in×25 ft) and capacities of 450 kW (600hp), the special-purpose lathes are used for applications such as railroad wheels, gun barrels and rolling-mill rolls. The cost of engine lathes ranges from about $ 2, 000 for bench types to over $100, 000 for larger units.

3. Lathe Operations

In a typical turning operation, the workpiece is clamped by any one of the workholding devices. Long and slender parts should be supported by a steady rest and follow rest placed on the bed; otherwise, the part will deflect under the cutting forces.[7] These rests are usually equipped with three adjustable fingers or rollers, which support the workpiece while allowing it to rotate freely. Steady rests are clamped directly on the ways of the lathe, whereas follow rests are clamped on the carriage and travel with it.

The cutting tool attached to the tool post, which is driven by the lead screw, removes material by traveling along the bed. A right-hand tool travels toward the headstock and a left-hand tool toward the tailstock. Facing operations are done by moving the tool radially with the cross-slide and clamping the carriage for better dimensional accuracy.

Form tools are used to produce various shapes on round workpieces by turning. The tool moves radially inward to machine the part. Machining by form cutting is not suitable for deep and narrow grooves or sharp corners because they may cause vibration and result in poor surface finish.[8] As a rule, ① the formed length should not be greater than about 2.5 times the minimum diameter of the part, ② the cutting speed should be reduced from turning settings, and ③ cutting fluids should be used.

The boring operation on a lathe is similar to turning. Boring is performed inside hollow workpieces or in a hole made previously by drilling or other means. Out-of-shape holes can be straightened by boring. The workpiece is held in a chuck or in some other suitable workholding device. Drilling can be performed on a lathe by mounting the drill bit in a drill chuck into the tailstock quill (a tubular shaft). The workpiece is placed in a workholder on the headstock, and quill is advanced by rotating the hand wheel.[9] Holes drilled in this manner may not be concentric because of the tendency for the drill to drift radially. The concentricity of the hole is improved by subsequently boring the drilled hole. Drilled holes may be reamed on lathes in a manner similar to drilling, thus improving hole tolerances.

Section Ⅱ New Words and Phrases

projection	*n.* 凸出，投影，投射
cycloidal	*adj.* 摆线的
cycloid	*n.* 摆线，轮转线
cycloidal profile	摆线轮廓

involute	*adj.* 渐开线
involute profile	渐开线轮廓
conjugate	*adj.* 共轭的
pinion	*n.* 小齿轮
dimension	*n.* 量纲，尺寸，维
mate	*n.* 啮合
engagement	*n.* 啮合
tangency	*n.* 接触，相切
pitch	*n.* 齿节，节距
intersect	*v.* 交叉
disposition	*n.* 排列，配置，布置
helical	*adj.* 螺旋状的
helical gear	螺旋齿轮，斜齿轮
spur	*n.* 刺
spur gear	正齿轮
worm	*n.* 蜗轮，蜗杆
bevel	*n.* 斜边和斜面
bevel gear	伞形齿轮
hourglass	*n.* 沙漏
V-belt	V 形带
splice	*n.* 连接，接合
groove	*n.* 沟，槽
tractive	*adj.* 牵引的，曳引的
clearance	*n.* 间隙，清算
lathe	*n.* 车床
pulley	*n.* 滑车，滑轮；（皮带）轮
manipulate	*vt.*（熟练地）操作，使用（机器等），操纵（人或市价、市场），利用，应付，假造
nodular	*adj.* 球状的；瘤状的
nodular cast iron	球墨铸铁
cross-section	横截面，断面，剖面图
carriage	*n.*（机械）车架
mount	*n.* 固定架 *vt.* 装上，设置，安放，固定，爬上，上演
assembly	*n.* 装配，组装，集合，集会，集结，汇编
tool post	刀架，刀座
apron	*n.* 挡板；刀座帷
compound rest	复式刀架；（车床）小刀架

swivel	*v.* 旋转
lead screw	导螺杆，丝杆
deflect	*v.* (使) 偏斜，(使) 偏转
headstock	*n.* 主轴箱，头架
hollow	穴，腔，窟窿 *adv.* 空的，中空的，空腹的，凹的
tubing	*n.* 装管，管道系统 *adj.* 管状的，管制的，制管的
spindle	*n.* 锭子，纺锤，轴，杆，心轴 *adj.* 锭子的，锭子似的
chuck	*n.* 卡盘
collet	*n.* 筒夹，夹头 *vt.* 镶进底座，装筒夹或夹头
tailstock	*n.* 尾架，尾座，顶针座
tailstock quill	尾架顶尖套筒，尾架顶心套筒
clamp	*n.* 夹子，夹具，夹钳 *vt.* 夹住，夹紧
workpiece	*n.* 工件，加工件
reamer	*n.* 钻孔器，刀，铰床
cylindrical	*adj.* 圆柱的
feed rod	分配杆，进给杆，进刀杆，光杆
friction clutch	摩擦离合器
nut	*n.* 螺母，坚果
thread	*n.* 螺纹，线，细丝，线索，思路
swing	*n.* 振幅，摆程，摇摆，摆动，秋千 *v.* 摇摆，摆动，回转，旋转
in	英寸（英制单位，1 in = 2. 54 cm）
ft	英尺（英制单位，1 ft = 30. 48 cm）
hp	马力（英制单位，1 hp = 745. 7 W）
diameter	*n.* 直径
bench lathe	台式车床
workbench	*n.* 工作台，成形台
precision-machine	精密加工
toolroom lathe	工具车，工具车床
close tolerance	紧公差，严格的容限
engine lathe	普通车床
gap bed lathe	马鞍式车床
slender	*adj.* 苗条的，微薄的，细长的
follow rest	跟刀架
form tool	样板刀；成形刀
groove	*n.* 凹槽 *vt.* 开槽于
cutting fluid	切削液；乳化切削油
workholder	*n.* 工件夹具，工件夹持装置

boring　　　　　　n. 钻（孔）
out-of-shape　　　形状不规则的

Section Ⅲ　Notes to Complex Sentences

[1] Gears are direct contact bodies, operating in pairs, that transmit motion and force from one rotating shaft to another, or from a shaft to a slide (rack), by means of successively engaging projections called teeth.

齿轮是直接接触、成对工作的实体。在称为齿的凸出物的连续啮合作用下，齿轮能将运动和力从一个旋转轴传递到另一个旋转轴，或从一个轴传递到一个滑块（齿条）。

operating in pairs 是分词短语，修饰前面的 Gears。

that 引导的从句，修饰前面的 Gears。

by means of 表示“借助”“通过”的意思。

[2] If an involute spur pinion is made of rubber and twists uniformly so that the ends rotates about the axis relative to one another, the elements of the teeth, initially straight and parallel to the axis would become helices.

如果一个渐开线直齿小齿轮是用橡皮制成的，能均匀扭曲，从而使一端以另一端为轴进行旋转，这样小齿轮上的齿开始时是直的并平行于传动轴，最后会变成螺旋形。

注意本句是虚拟语句，

were made of...由……组成。

so that 引导结果状语从句。

parallel to...：平行于……。

[3] Sufficient clearance must be provided at the bottom of the groove to prevent the belt from bottoming as it becomes narrower from wear.

在带轮槽的底部需要留有足够的间隙，以保证 V 形带不接触带轮槽的底部，否则会因磨损而变得越来越窄。

at the bottom of...：在……的底部。

prevent...from bottoming：防止触底。

[4] The cutting tool is mounted on the tool post, usually with a compound rest that swivels for tool positioning and adjustment. The cross-slide moves radially in and out, controlling the radial position of the cutting tool in operations such as facing.

刀具安装在刀架上，通常采用旋转复合刀架，以便刀具的定位和调整。横向滑板可以径向移进、移出，以控制切削加工中（如车端面）刀具的径向位置。

[5] It rotates during the operation of the lathe and provides movement to the carriage and the cross-slide by means of gears, a friction clutch, and a keyway along the length of the rod.

进给杆在车床操作时可以旋转，然后靠齿轮、摩擦离合器和长杆键槽给机架和横向滑板提供运动量。

by means of 意为“用，依靠”。

E. g., thoughts are expressed by means of words.

思想通过语言来表达。

[6] Bench lathes are placed on a workbench. They have low power and are usually operated by hand feed, which are used for precision-machine small workpieces.

台式车床通常放在工作台上，其功率低，通常手动操作进刀，用来精密加工小工件。

[7] Long and slender parts should be supported by a steady rest and follow rest placed on the bed; otherwise, the part will deflect under the cutting forces.

细长零件必须由一个稳定的支撑架和安放在机床上的跟刀架支撑；否则，零件就会在切削力的作用下偏转。

follow rest 意为“跟刀架”。

[8] The tool moves radially inward to machine the part. Machining by the form of cutting is not suitable for deep and narrow grooves or sharp corners because they may cause vibration and result in poor surface finish.

成形刀径向移动加工零件。成形刀不适合深且窄的凹槽加工或是锐角转角加工，由于它们加工时会振动，从而导致完成的表面质量较差。

[9] The workpiece is placed in a workholder on the headstock and quill is advanced by rotating the hand wheel.

加工件用工件夹具固定在主轴箱上，尾架顶尖套筒靠手轮旋转向前运动。

Section Ⅳ Exercises

Translate the following paragraphs into Chinese.

A gear having tooth elements that are straight and parallel to its axis is known as a spur gear. A spur pair can be used to connect parallel shafts only. Parallel shafts, however, can also be connected by gears of another type and a spur gear can be mated with a gear of a different type. Helical gears have certain advantages, for example, when connecting parallel shafts, they have a higher load-carrying capacity than spur gears with the same tooth numbers and cut with the same cutter. Helical gears can also be used to connect nonparallel and non-intersecting shafts at any angle to one another. Ninety degrees is the most common angle at which such gears are used.

Worm gears provide the simplest means of obtaining large ratios in a single pair. They are usually less efficient than parallel shaft gears. However, because of their similarity, the efficiency of a worm and gear depends on the same factors as the efficiency of a screw.

(1) Although simple and versatile, an engine lathe requires a skilled machinist because all controls are manipulated by hand.

(2) Headstocks have a hollow spindle to which workholding devices, such as chucks and collets are attached and long bars or tubing can be fed through for various turning operations.

(3) Drilling can be performed on a lathe by mounting the drill bit in a drill chuck into the tailstock quill (a tubular shaft).

Section V Supplementary Reading

1. Chain

The first chain-driven or "safety" bicycle appeared in 1874 and chains were used for driving the rear wheels on early automobiles. Today, as the result of modern design and production methods, chain drives that are much superior to their prototypes are available and these have contributed to the development of efficient agricultural machinery, well-drilling equipment, mining and construction machinery. Since about 1930, chain drives have become increasingly popular, especially for power saws, motorcycle, and escalators, etc. Fig. 5-5-1 shows a chain drive.

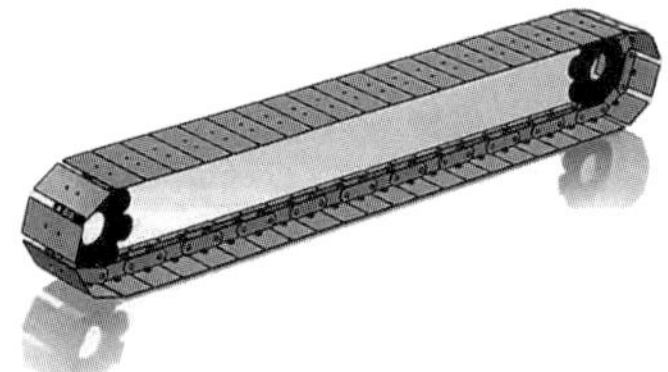

Fig. 5-5-1 Chain drive

There are at least six types of power-transmission chains, three of these will be covered in this article, namely the roller chain, the inverted tooth, or silent chain and the bead chain. The essential elements in a roller chain drive are a chain with side plates, pins, bushings (sleeves), rollers and two or more sprocket wheels with teeth that look like gear teeth. Roller chains are assembled from pin links and roller links. A pin link consists of two side plates connected by two pins inserted into holes in the side plates. The pins fit tightly into the holes, forming what is known as a press fit. A roller link consists of two side plates connected by two press-fitted bushings, on which two hardened steel rollers are free to rotate. When assembled, the pins are a free fit in the bushings and rotate slightly relative to the bushings when the chain goes on and leaves a sprocket. Fig. 5-5-2 shows different types of chain drives.

Fig. 5-5-2 Different types of chain drives

Standard roller chains are available in single strands or in multiple strands. In the latter type, two or more chains are joined by common pins that keep the rollers in the separate strands in proper alignment. The speed ratio for a single drive should be limited to about 10 ∶ 1, the preferred shaft center distance is 30−35 times the distance between the rollers and chain speeds greater than about Boom/min(2, 500 ft/min) are not recommended. If several parallel shafts are to be driven without slip from a single shaft, roller chains are particularly well suited.

An inverted tooth or silent chain is essentially an assemblage of gear racks, each with two teeth pivotally connected to form a closed chain with the teeth on the inside and meshing with conjugate teeth on the sprocket wheels. The links are pin-connected flat steel plates which usually have straight-sided teeth with an included angle of 60°. As many links are necessary to transmit the power and are connected side by side. Compared with roller-chain drives, silent-chain drives are quieter, operate successfully at higher speeds and can transmit more load for the same width. Some automobiles have silent-chain camshaft drives.

Bead chains provide an inexpensive and versatile means for connecting parallel or nonparallel shafts when the speed and power transmitted are low. The sprocket wheels contain hemispherical or conical recesses into which the beads fit. The chains look like key chains and are available in plain carbon, stainless steel and also in the form of solid plastic beads molded on a cord. Bead chains are used on computers, air conditioners, television tuners, and Venetian blinds. The sprockets may be steel, die-cast zinc, aluminum or molded nylon.

2. Milling

Milling (Fig. 5−5−3) is a machining process for removing material by relative motion between a workpiece and a rotating cutter having multiple cutting edges. In some applications, the workpiece is held stationary while the rotating cutter is moved to past it at a given feed rate (traversed). In other applications, both the workpiece and cutter are moved in relation to each other and in relation to the milling machine. More frequently, however, the workpiece is advanced at a relatively low rate of movement or feed to a milling cutter rotating at comparatively high speed, with the cutter axis remaining in a fixed position. A characteristic feature of the milling process is that each milling cutter tooth takes its share of the stock in the form of small individual chips. Milling operations are performed on many different machines.

Since both the workpiece and cutter can be moved relative to one another independently or in combination, a wide variety of operations can be performed by milling. Applications include the production of flat or contoured surfaces, slots, grooves, recesses, threads and other configurations.

Milling is one of the most universal, yet complicated machining methods. The process has more variations in the kinds of machines used, workpiece movements and types of tooling than any other basic machining method. Important advantages of removing material by means of milling include high stock removal rates, the capability of producing relatively smooth surface finishes and the wide variety of cutting tools that are available. Cutting edges of the tools can be shaped to form any complex surface.

Fig. 5-5-3 Milling

The major milling methods are peripheral and face milling; in addition, a number of related methods exist that are variations of these two methods, depending upon the type of workpiece or cutter.

Exercises

1. Write T(True) or F(False) beside the following statements about the text.

(1) ________The most common lathe was originally called an engine lathe because it was powered with overhead pulleys and belts from nearby engines.

(2) ________The top portion of the bed has two ways, with various cross-sections that are punched and hardened for water proof and dimensional accuracy during use.

(3) ________The cutting tool is mounted on the tool post, usually, with a compound rest that swivels for tool positioning and adjustment.

(4) ________The headstock is closed to the bed and is equipped with engines, pulleys and V-belts that supply steam and electricity to the spindle at various rotational speeds.

(5) ________ Drills and reamers can be mounted on the tailstock quill (a hollow cylindrical part with a tapered hole) to drill axial holes in the workpiece.

(6) ________Engine lathes are only in several types of the sizes and are used for a variety of turning operations. In gap bed lathes, a section of the bed at the back of the headstock can be removed to fix and tighten larger-diameter workpieces.

(7) ________Machining by form cutting is suitable for deep and narrow grooves or sharp corners because they must cause vibration and result in smooth and good surface finish.

(8) ________The boring operation on a lathe is similar to turning. Boring is performed inside hollow workpieces or in a hole made previously by drilling or other means. Out-of-shape holes can be straightened by boring.

2. Choose the best answer.

(1) Helical gears have certain advantages, for example, when connecting __________, they have a higher load-carrying capacity than spur gears with the same tooth numbers and cut with the same cutter.

A. parallel engine B. parallel spindle C. parallel spring D. parallel shaft

(2) They are usually less efficient than parallel shaft gears, however, because of an additional __________ movement along the teeth.

A. running B. sliding C. circling D. spiral

(3) Since about 1930, __________ have become increasingly popular, especially for power saws, motorcycle, and escalators, etc.

A. manual drives B. active drives C. chain drives D. electric drives

(4) The __________ elements in a roller-chain drive are a chain with side plates, pins, bushings (sleeves), rollers and two or more sprocket wheels with teeth that look like gear teeth.

A. additional B. essential C. extra D. complementary

(5) The __________ for a single drive should be limited to about 10: 1, the preferred shaft center distance is from 30−35 times the distance between the rollers and chain speeds greater than about 800 m/min (2, 500 ft/min) are not recommended.

A. speed ratio B. speed classification

C. speed limitation D. speed control

3. Fill in the blanks with words or expressions according to the text.

(1) Although simple and versatile, __________ requires a skilled machinist because all controls __________.

(2) The top portion of the bed has two ways, with various __________ that are hardened and machined for __________ during use.

(3) The apron is equipped with mechanisms for both __________ of the carriage and the cross-slide by means of the __________.

(4) In gap bed lathes, a section of the bed in front of the __________ can be removed to accommodate larger-diameter __________.

(5) __________ are done by moving the tool radially with the cross-slide and clamping the carriage for better __________.

(6) __________ may be reamed on lathes in a manner similar to drilling, thus improving __________.

4. Answer the following questions according to the text.

(1) 装在机床床头箱主轴上的卡盘通常是什么类型的卡盘？什么情况下需要用四爪卡盘？

(2) 车床是否总是以一种速度运转？如果不是，那么在什么情况下需要采用不同的转速？

（3）是否所有的工件都必须固定在机床的两个顶尖之间？如果不是，什么情况下不需要用尾架顶尖，怎样处理？

5. Translate the following paragraphs into Chinese.

(1) The first chain-driven or "safety" bicycle appeared in 1874 and chains were used for driving the rear wheels on early automobiles. Today, as the result of modern design and production methods, chain drives that are much superior to their prototypes are available and these have contributed greastly to the development of efficient agricultural machinery, well-drilling equipment, mining and construction machinery.

(2) Bead chains provide an inexpensive and versatile means for connecting parallel or nonparallel shafts when the speed and power transmitted are low. The sprocket wheels contain hemispherical or conical recesses into which the beads fit. The chains look like key chains and are available in plain carbon, stainless steel and also in the form of solid plastic beads molded on a cord. Bead chains are used on computers, air conditioners, television tuners, and Venetian blinds. The sprockets may be steel, die-cast zinc, aluminum or molded nylon.

(3) Milling is one of the most universal, yet complicated machining methods. The process has more variations in the kinds of machines used, workpiece movements and types of tooling than any other basic machining method. Important advantages of removing material by means of milling include high stock removal rates, the capability of producing relatively smooth surface finishes and the wide variety of cutting tools that are available. Cutting edges of the tools can be shaped to form any complex surface.

单元评价

通过本单元的学习，使学生掌握了零件制造与普通加工相关的英语词汇及常用表达语句，并具备运用英语检索零件制造与普通加工相关信息及使用英语描述零件制造与普通加工相关内容的能力，同时培养学生实事求是、统筹兼顾、专注的工匠精神。

单元 6

液压与气动回路设计

【学习目标】

知识目标：

1. 熟悉液压与气动回路设计的相关英语词汇。
2. 掌握与人交流液压与气动回路设计时常用的英语表达语句。

技能目标：

1. 具备液压与气动回路设计英语词汇寻读和跟读技巧，能够运用英语检索相关信息。
2. 具备使用英语描述液压与气动回路设计相关信息和过程的能力。

素养目标：

1. 培养学生踏实严谨、精益求精的治学态度。
2. 培养学生爱岗敬业、团结协作的工作作风。
3. 培养学生制造强国、科技强国的使命担当意识。

Section I Texts

The word "hydraulics" generally refers to the power produced by moving liquids. Modern hydraulics is defined as the use of the confined liquid to transmit the power, multiply force, or produce motion. Though hydraulic power in the form of water wheels and other simple devices has been in use for centuries, the principles of hydraulics weren't formulated into scientific law until the 17th century. It was then that French philosopher Blaise Pascal discovered that the liquids cannot be compressed.[1] He discovered a law which states that the pressure applied on a confined fluid is transmitted in all directions with the equal force on equal areas.

Hydraulic systems contain the following key components.

(1) Fluid—can be almost any liquid. The most common hydraulic fluids contain specially compounded petroleum oils that lubricate and protect the system from corrosion.[2]

(2) Reservoir or tank—acts as a storehouse for the fluid and a heat dissipater.

(3) Hydraulic pump—converts the mechanical energy into hydraulic energy by forcing hydraulic fluid, under pressure, from the reservoir into the system.

(4) Fluid lines or pipes—transport the fluid to and from the pump through the hydraulic

system. These lines can be rigid metal tubes or flexible hose assemblies. Fluid lines can transport the fluid under the pressure or vacuum (suction).

(5) Hydraulic valves—control the pressure, direction and flow rate of the hydraulic fluid.

(6) Actuator—converts hydraulic energy into mechanical energy to do work. Actuators usually take the form of hydraulic cylinders. Hydraulic cylinders are used on agricultural, construction, and industrial equipment.

In actual hydraulic systems, Pascal's law defines the basis of the results which are obtained from the system. Thus, a pump moves the liquid in the system. The intake of the pump is connected to the reservoir or tank. Atmospheric pressure, pressing on the liquid in the reservoir, forces the liquid into the pump. When the pump operates, it forces liquid from the tank into the discharge pipe at a suitable pressure. The flow of the pressurized liquid discharged by the pump is controlled by the valves. Three control functions are used in most hydraulic systems: ①control of the liquid pressure, ②control of the liquid flow rate, and ③control of the direction of flow of the liquid.

Hydraulic drives are used in preference to mechanical systems when ① power is to be transmitted between points too far apart for chains or belts; ② high torque at a low speed is required; ③a very compact unit is needed; ④a smooth transmission, free of vibration, is required; ⑤easy control of speed and direction is necessary; and ⑥output speed is varied stepless.

Fig. 6-1-1 gives a diagrammatic presentation of the components of a hydraulic system installation. Electrically driven oil pressure pumps establish an oil flow for energy transmission, which is fed to hydraulic motors or hydraulic cylinders, converting it into the mechanical energy. The control of the oil flow is by means of valves. The pressurized oil flow produces linear or rotary mechanical motion. The kinetic energy of the oil flow is comparatively low, and, therefore, the term hydrostatic driver is sometimes used. There is little constructional difference between hydraulic motors and pumps. Any pump may be used as a motor. The quantity of oil flowing at any given time may be varied by means of the regulating valves or the use of variable-delivery pumps. Fig. 6-1-2 shows hydrautic motor drive system.

Fig. 6-1-1 Hydraulic system

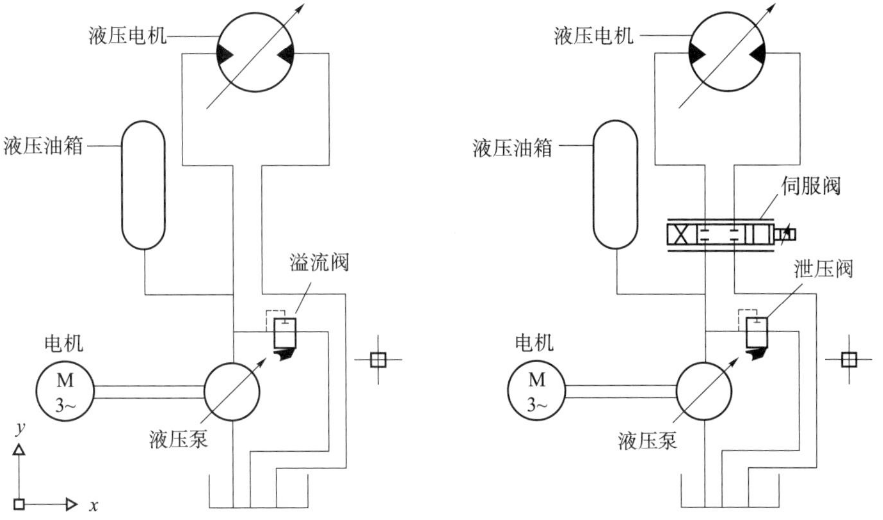

Fig. 6-1-2　Hydrautic motor drive system

Hydraulic motor drives offer a great many advantages.[3] One of them is that it can give infinitely-variable speed control over wide ranges. In addition, they can change the direction of drive as easily as they can vary the speed. As in many other types of machines, many complex mechanical linkages can be simplified or even wholly eliminated by the use of hydraulics.

The flexibility and resilience of hydraulic motor drive is another great virtue of this form of drives. Apart from the smoothness of the operation thus obtained, a great improvement is usually found in the surface finish on the work and the tool can make heavier cuts without detriment and will last considerably longer without regrinding.[4]

Section Ⅱ　New Words and Phrases

hydraulics	*n.* 液压技术，水力学
confine	*vt.* 封闭
transmit	*v.* 传送
multiply	*v.* 乘，增加，放大
reservoir	*n.*（油）箱，水库
hose	*n.* 软管，水龙头
valve	*n.* 阀门
actuator	*n.* 执行器
intake	*n.* 进（气、水）口
discharge	*v.* 卸载，卸货
torque	*n.* 力矩
vibration	*n.* 振动

steplessly	*adv.* 无级地
linear	*adj.* 线性的
rotary	*adj.* 旋转的
diagrammatic	*adj.* 图表的，概略的，diagram 图
oil pressure pump	油泵
hydraulic motor	液压电机
hydraulic cylinder	油缸
kinetic energy	动能
hydrostatic driver	静压传动
variable-delivery pump	变量泵
by no means	决不
self-contained	独立的，配套的，整体的
regulating valve	调压阀
stimulate	*vt.* 促进，激励
resilience	*n.* 跳回，恢复力，回弹
virtue	*n.* 优点，效力，功能
finish	*n.* 光洁度
work	*n.* 工件
cut	*n.* 进刀量
detriment	*n.* 损害，不利
regrind	*v.* 重磨，grind 磨（刀）

Section Ⅲ Notes to Complex Sentences

[1] It was then that French philosopher Blaise Pascal discovered that liquids cannot be compressed.

It was then that…是强调句型；it 是形式主语；then 是表语；that 引导的从句作逻辑主语，强调 then。第二个 that 引导的从句作 discovered 的宾语。

全句可译为“就在那时，法国哲学家巴斯卡发现液体是不能被压缩的”。

[2] The most common hydraulic fluids contain specially compounded petroleum oils that lubricate and protect the system from corrosion.

最常见的液压流体含有特制的石油化合油，起到润滑作用并保护系统免受腐蚀。

contain：含有；that 引导的定语从句修饰 oils，翻译时可转换成状语。

[3] Hydraulic motor drives offer a great many advantages.

液压电机驱动具有许多优点。

这里，drive 是名词，offer 是谓语，a great many 意为“许多”。

[4] ...a great improvement is usually found in the surface finish on the work and the tool can make heavier cuts without detriment and will last considerably longer without regrinding.

……通常可见工件的表面粗糙度得到大幅改善，刀具进刀量可以加大而不致损伤，并且持续更长时间无须重磨。

surface finish：表面粗糙度；tool：刀具；cuts：切削量；last：持续。

Section Ⅳ Exercises

Translate the following paragraph into Chinese.

Compressors used in the most common hydraulic fluids contain specially compounded petroleum oils that lubricate and protect the system from the corrosion. Compressors are used in petrochemical plants to raise the static pressure of air and process gases to levels required to overcome pipe friction, to affect a certain reaction at the point of final delivery, or to impart desired thermodynamic properties to the medium compressed. These compressors come in a variety of sizes, types and models, each of which fulfills a given need and is likely to represent the optimum configuration for a given set of requirements. Selection of the compressor types must, therefore, be preceded by a comparison between the service requirements and the compressor capabilities. This initial comparison will generally lead to a review of the economies of space, installing cost, operating cost and maintenance requirements of the competing types. If the superiority of one compressor type or model over a competing offer is not obvious, a more detailed analysis may be justified.

Section Ⅴ Supplementary Reading

Mechanism Systems

While some simple machines consist of only one kind of mechanisms, in most cases, using only one simple mechanism is not enough to perform the required mechanical actions in a machine. Two working links (or output links) are needed to shape a flat surface. They are the sliding block with the shaping tool (cutter) and the worktable holding the workpiece. Carrying the cutter, the sliding block moves back and forth to perform the cutting motion and the stroke of this motion is adjustable to fit the size of the workpiece. The workable moves intermittently to provide the feeding action while the sliding block moves back. The amount of feed is also adjustable. Such a working process needs several simple mechanisms working together in a machine and form a mechanism system. Another example of a mechanism system is the well-known internal combustion engine, which consists of a crank and slider mechanism, cam mechanisms and a gear mechanism. The crank and slider mechanism convert the back and forth movement of the piston into the rotation of the crankshaft. The gear mechanism and cam mechanisms control the movements of the valves exactly and ensure the synchronized operation of the whole engine. According to system theory, a machine can be seen as a system of the mechanisms and a mechanism is a sub-system of a machine. Hence, the design of a machine is the design of a mechanism system.

The quality, performance and compatibility of a mechanical product depend mainly on its design. Any error, defect or carelessness in design may result in considerable extra cost in manufacture or even the failure of the product. The importance of the design is obvious here.

Exercises

1. Write T(True) or F(False) beside the following statements about the text.

(1) ________ Though hydraulic power in the form of water wheels and other simple devices has been in use for centuries, the principles of hydraulics weren't formulated into scientific law until the 17th century.

(2) ________ The intake of the pump is connected to the reservoir or tank. Atmospheric pressure, pressing on the air in the tank, forces the liquid out of the pump.

(3) ________ Electrically driven oil pressure pumps establish an oil flow for energy supply, which is fed to hydraulic gears or mechanical cylinders, converting it into the mechanical energy.

(4) ________ In addition, they can change the direction of drive as possible as they can vary the speed. As in many other types of machine, many complex mechanical linkages can be changed or even wholly converted by the use of mechanical settings.

(5) ________ Apart from the smoothness of the operation thus obtained, a great improvement is usually found in the surface finish on the work and the tool can make heavier cuts without detriment and will last considerably longer without regrinding.

(6) ________ These compressors come in a variety of sizes, types and models, each of which fulfills a given need and is likely to represent the optimum configuration for a given set of requirements.

(7) ________ Carrying the cutter, the sliding block moves in order to perform the cutting motion and the stroke of this motion is adapted to fit the shape and form of the workpiece.

(8) ________ Any error, defect or carelessness in design may result in considerable extra cost in transportation or even the output of the product. The importance of the production is obvious here.

2. Choose the best answer.

(1) The most common hydraulic fluids contain specially compounded petroleum oils that lubricate and protect the system from________.

A. production　　B. process　　C. corrosion　　D. sinking

(2) In actual ________ systems, Pascal's law defines the basis of the results which are obtained from the system.

A. steam　B. hydraulic　C. mechanical　D. electrical

(3) The__________ of the oil flow is comparatively low, and, therefore, the term hydrostatic driver is sometimes used.

A. kinetic power　B. dynamic force　C. solar power　D. electric power

(4) As in many other types of machine, many complex__________can be simplified or even wholly eliminated by the use of hydraulics.

A. electric linkage　B. hydraulic linkage　C. intelligent linkage　D. mechanical linkages

(5) These compressors come in a variety of sizes, types and models, each of which fulfills a given need and is likely to represent the optimum__________for a given set of requirements.

A. settings　B. backgrounds　C. restorations　D. configuration

3. Fill in the blanks with words or expressions according to the text.

(1) He discovered a law which states that the__________ applied on a confined __________ is transmitted in all directions with the equal force on equal areas.

(2) When the__________operates, it forces liquid from the tank into the __________at a suitable pressure.

(3) Electrically driven oil__________establish an oil flow for __________, which is fed to hydraulic motors or hydraulic cylinders, converting it into the mechanical energy.

(4) As in many other types of machine, many complex __________can be simplified or even wholly eliminated by the use of__________.

(5) Such a__________needs several simple__________working together in a machine and form a mechanism system.

(6) According to system theory, a machine can be seen as a system of the mechanisms and a mechanism is a __________of a machine. Hence, the design of a machine is the design of a __________.

4. Answer the following questions according to the text.

(1) 什么是液压系统？液压系统的组成部分有哪些？

(2) 液压电机速度控制的方法是什么？

(3) 在什么情况下采用液压系统？液压系统的优缺点是什么？

5. Translate the following paragraphs into Chinese.

(1) Compressors used in the most common hydraulic fluids contain specially compounded petroleum oils that lubricate and protect the system from the corrosion. Compressors are used in petrochemical plants to raise the static pressure of air and process gases to levels required to overcome pipe friction, to affect a certain reaction at the point of final delivery, or to impart desired thermodynamic properties to the medium compressed.

(2) The crank and slider mechanism converts the back and forth movement of the piston into the rotation of the crankshaft. The gear mechanism and cam mechanisms control the movements of

the valves exactly and ensure the synchronized operation of the whole engine.

(3) The quality, performance and compatibility of a mechanical product depend mainly on its design. Any error, defect or carelessness in design may result in considerable extra cost in manufacture or even the failure of the product. The importance of the design is obvious here.

单元评价

通过本单元的学习，使学生掌握液压与气动回路设计相关的英语词汇及常用表达语句，并具备运用英语检索液压与气动回路设计相关信息及使用英语描述液压与气动回路设计相关内容的能力，同时培养学生踏实严谨、精益求精的治学态度和制造强国、科技强国的使命担当意识。

单元 7

机械制造工艺及仿真

【学习目标】

知识目标：

1. 熟悉机械制造工艺及仿真的相关英语词汇。
2. 掌握与人交流机械制造工艺及仿真时常用的英语表达语句。

技能目标：

1. 具备机械制造工艺及仿真英语词汇寻读和跟读技巧，能够运用英语检索相关信息。
2. 具备使用英语描述机械制造工艺及仿真相关信息和过程的能力。

素养目标：

1. 培养学生不断上进、吃苦耐劳的作风和爱岗敬业的精神。
2. 培养学生较强的语言、文字表达能力和社会沟通能力。
3. 培养学生制造强国、科技强国的使命担当意识。

Section Ⅰ Texts

Text 1

In order to make a manufacturing operation be efficient, all of its diverse activities must be planned. This activity has traditionally been done by process planners. Process planning is concerned with the selecting methods of production: tooling, fixtures, machinery, sequence of operations, and assembly.[1]

The sequence of the processes and operations to be performed, the machines to be used, the standard time for each operation and similar information are all documented on a routing sheet (Table. 7-1-1). When done manually, this task is highly labor-intensive and time-consuming and relies heavily on the experience of the process planner. A current trend in routing sheets is to store the relevant data in computers and affix a bar code (or other identification) to the part. The production data can then be reviewed at a dedicated monitor.

Table 7-1-1 A Typical Routing Sheet

Routing Sheet

Customer: S Name: Midwest Valve Co. Part Name: Valve body

Quantity: 15 Part No. : 302

Operation No.	Description of operation	Machine
10	Inspect forging check hardness	Rockwell tester
20	Rough machine flanges	Lathe No. 5
30	Finish machine flanges	Lathe No. 5
40	Bore and counter bore hole	Boring mill No. 1
50	Turn internal grooves	Boring mill No. 1
60	Drill and tap holes	Drill press No. 2
70	Grind flange end faces	Grinder No. 2
80	Grind bore	Internal grinder No. 1
90	Clean	Vapor degreaser
100	Inspect	Ultrasonic tester

Computer-aided process planning (CAPP) accomplishes this complex task of the process planning by viewing the total operation as an integrated system, so that the individual operations and steps involved in making each part are coordinated with others and are performed efficiently and reliably.[2] Thus, computer-aided process planning is an important adjunct to CAD/CAM.

Although CAPP requires extensive software and good coordination with CAD/CAM, it is a powerful tool for efficiently planning and scheduling manufacturing operations. CAPP is particularly effective in small-volume, high-variety parts production requiring machining, forming, and assembly operations.

1. Element of CAPP Systems

There are two types of computer-aided process planning systems, variant and generative process planning.

In the variant system (also called the derivative system), the computer files contain a standard process plan for the part to be manufactured. The search for a standard plan is made in the database by a code number for the part; the plan is based on its shape and its manufacturing characteristics. The standard plan is retrieved, displayed for review and printed as a routing sheet.

The process plan includes information such as the types of tools and machines to be used, the sequence of manufacturing operations to be performed, the speeds, the feeds, the time required for each sequence and so on. Minor modifications of an existing process plan (which are usually necessary) can also be made. If the standard plan for a particular part is not in the computer files, a plan that is close to it, with a similar code number and an existing routing sheet, is retrieved.[3] If a routing sheet does not exist, one is made for the part and stored in computer memory.

In the generative system, a process plan is automatically generated on the basis of the same logical procedures that would be followed by a traditional process planner in making that particular part.[4] Such a system is complex, however, because it must contain comprehensive and detailed knowledge of the part shape and dimensions, of the process capabilities, about the selection of manufacturing methods, machinery, and tools and of the sequence of the operations to be performed.

The generative system is capable of creating a new plan instead of having to use and modify an existing plan. Although, currently, it is used less commonly than is variant system, this system has such advantages as ① flexibility and consistency for process planning for new parts and ② higher overall planning quality, because of the capability of the decision logic in the system to optimize the planning and to utilize up-to-date manufacturing technology.[5]

2. Advantages of CAPP Systems

The advantages of CAPP systems over traditional process planning methods include the following.

(1) The standardization of the process plans improves the productivity of the process planners, reduces lead times, reduces planning costs and improves the consistency of the product quality and reliability.

(2) Process plans can be prepared for the parts having similar shapes and features, and they can be retrieved easily to produce new parts.

(3) Process plans can be modified to suit the specific needs.

(4) Routing sheets can be prepared more quickly. Compared to the traditional handwritten routing sheets, the computer printouts are neater and much more legible.

(5) Other functions, such as the cost estimating and work standards, can be incorporated into CAPP.

Exercises

1. Write T(True) or F(False) beside the following statements about the text.

(1) ________ Process planning is concerned with the selecting methods of production: tooling, fixtures, machinery, sequence of operations, and assembly.

(2) ________ The sequence of the processes and operations to be performed, the machines to be used, the standard time for each operation and similar information are not all documented on the same routing sheet.

(3) ________ When done automatically, this task is highly labor-saving and time-saving and relies lightly on the experience of the process planner.

(4) ________ Computer-aided process planning (CAPP) accomplishes this complex task

of the process planning by viewing the total operation as an integrated system, so that the individual operations and steps involved in making each part are coordinated with others and are performed efficiently and reliably.

(5) __________Although CAPP requires only a few amount of software and good coordination with CAD/CAM, it is an ordinary tool for efficiently planning and scheduling manufacturing production.

(6) __________The search for a standard plan is made in the database by a code number for the part; the plan is based on the production and the manufacturing source.

(7) __________Such a system is complex, however, because it must contain comprehensive and detailed knowledge of the part shape and dimensions, of the process capabilities, about the selection of manufacturing methods, machinery, and tools and of the sequence of the operations to be performed.

(8) __________The standardization of the process plans decreases the productivity of the process planners, increases lead times, reduces planning costs and the consistency of the product quality and reliability.

2. Choose the best answer.

(1) Process planning is concerned with the selecting methods of __________: tooling, fixtures, machinery, sequence of operations, and assembly.

A. production B. productivity C. producer D. productive

(2) A current trend in routing sheets is to store the relevant data in computers and affix a __________ (or other identification) to the part.

A. bar number B. bar identification C. bar weight D. bar code

(3) CAPP is particularly __________ in small-volume, high-variety parts production requiring machining, forming, and assembly operations.

A. influencing B. effective C. influential D. effecting

(4) In the generative system, a __________ is automatically generated on the basis of the same logical procedures that would be followed by a traditional process planner in making that particular part.

A. process plan B. process method

C. process characteristics D. process impression

(5) The generative system is capable of creating a new plan instead of having to use and modify an__________ plan.

A. existed B. existent C. exiting D. exist

3. Fill in the blanks with words or expressions according to the text.

(1) When done manually, this task is highly __________ and __________ and relies heavily on the experience of the process planner.

(2) A current trend in __________ is to store the relevant data in computers and affix a __________ (or other identification) to the part.

(3) The search for a __________ is made in the database by a code number for the part; the plan is based on its shape and its __________.

(4) __________ of an existing __________ (which are usually necessary) can also be made.

(5) The standardization of the process plans improves the __________ of the process planners, reduces lead times, reduces planning costs and improves the consistency of the __________.

(6) Compared to the traditional handwritten __________, the __________ are neater and much more legible.

4. Answer the following questions according to the text.

(1) 工序设计主要关注什么？什么是计算机辅助工艺设计？

(2) 什么是 CAPP 系统，它的构成要素有什么？

(3) CAPP 系统的优点是什么？

5. Translate the following paragraphs into Chinese.

(1) In order to make a manufacturing operation be efficient, all of its diverse activities must be planned. This activity has traditionally been done by process planners. Process planning is concerned with the selecting methods of production: tooling, fixtures, machinery, sequence of operations, and assembly.

(2) Computer-aided process planning(CAPP) accomplishes this complex task of the process planning by viewing the total operation as an integrated system, so that the individual operations and steps involved in making each part are coordinated with others and are performed efficiently and reliably. Thus, computer-aided process planning is an important adjunct to CAD and CAM.

(3) The process plan includes information such as the types of tools and machines to be used, the sequence of manufacturing operations to be performed, the speeds, the feeds, the time required for each sequence and so on. Minor modifications of an existing process plan (which are usually necessary) can also be made. If the standard plan for a particular part is not in the computer files, a plan that is close to it, with a similar code number and an existing routing sheet, is retrieved. If a routing sheet does not exist, one is made for the part and stored in computer memory.

Text 2

A flexible manufacturing system(FMS) (Fig. 7-1-1) integrates all major elements of the manufacturing to a highly automated system. Firstly, utilized in the late 1960s, FMS consists of a number of manufacturing cells, each containing an industrial robot (serving several CNC machines) and an automated material-handling system, all interfaced with a central computer. Different computer instructions for the manufacturing process can be downloaded for each successive part passing through the workstation.

Fig. 7-1-1 Flexible manufacturing unit system

This system is highly automated and is capable of optimizing each step of the total manufacturing operation. These steps may involve one or more processes and operations (such as machining, grinding, cutting, forming, powder metallurgy, heat treating, and finishing), as well as handling of raw materials, inspection, and assembly. The most common applications of FMS to date have been in machining and assembly operations. A variety of FMS technology is available from the machine-tool manufactures.[6]

Flexible manufacturing systems represent the highest level of efficiency, sophistication and productivity that has been achieved in the manufacturing plants. The flexibility of FMS is such that it can handle a variety of part configurations and produce them in any order.

FMS can be regarded as a system which combines the benefits of two other systems, ① the highly productive but inflexible transfer lines, and ② job-shop production, which can produce large product variety on stand-alone machines, but inefficient.[7]

The basic elements of a flexible manufacturing system are ① workstations, ② automated handling and transport of materials and parts, and ③ control systems. The workstations are arranged to yield the greatest efficiency in production, with an orderly flow of materials, parts and products through the system.

The types of machines in workstations depend on the type of production. For machining operations, they usually consist of a variety of 3-5 axis machining centers, CNC lathes, milling machines, drill presses and grinders. Other equipment, such as that for automated inspection, assembly and cleaning are also included.

Other types of operations suitable for FMS include sheet metal forming, punching and shearing and forging; they incorporate furnaces, forging machines, trimming presses, heat-treating facilities and cleaning equipment.

Because of the flexibility of FMS, material-handling, storage and retrieval systems are very important. Material handling is controlled by a central computer and performed by automated guided vehicles, conveyors, and various transfer mechanisms. The system is capable of transporting raw materials, blanks and parts in various stages of completion to any machine (in random order) and at any time.[8] Prismatic parts are usually moved on specially designed pallets. Parts having rotational symmetry (such as those for turning operations) are usually moved by mechanical devices and robots.

The computer control system of FMS is its brains and includes various software and hardware. This sub-system controls the machinery and equipment in workstations and the transporting of raw materials, blanks and parts in various stages of completion from machine to machine. It also stores data and provides the communication terminals that display the data visually. See Fig. 7-1-2.

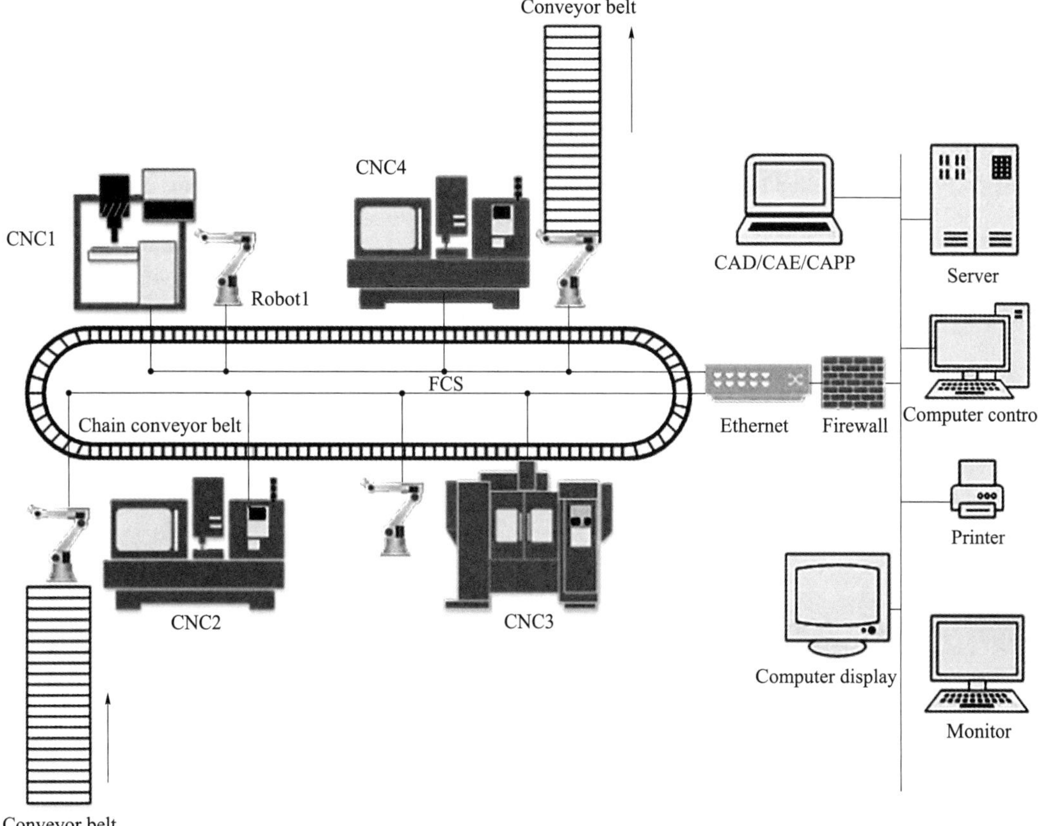

Fig. 7-1-2　Process transfer of flexible manufacturing unit system

Because FMS involves a major capital investment, efficient machine utilization is essential: machines must not stand idle. Consequently, the proper scheduling and the process planning are crucial. Scheduling for FMS is dynamic, unlike that in job shops, where a relatively rigid schedule is followed to perform a set of operations. The scheduling system for FMS specifies the types of operations to be performed on each part and it identifies the machines or manufacturing cells to be used. Dynamic scheduling is capable of responding to quick changes in product type and so is responsive to real-time decisions.[9]

Because of the flexibility in FMS, no setup time is wasted in switching between manufacturing operations, the system is capable of performing different operations in different orders and on different machines. However, the characteristics, performance and reliability of each unit in the system must be checked to ensure that parts moving from workstation to workstation are of the acceptable quality and dimensional accuracy.[10]

FMS installations are very capital-intensive, typically starting at well over $1 million. Consequently, a thorough cost-benefit analysis must be conducted before a final decision is made. This analysis should include such factors as the cost of capital, of energy, of materials and of labor, the expected markets for the products to be manufactured and any anticipated fluctuations in market demand and product type.[11] An additional factor is the time and effort required for installing and debugging the system.

Typically, an FMS system can take two to five years to install and at least six months to debug. Although FMS requires few, if any, machine operators, the personnel in charge of the total operation must be trained and highly skilled. These personnel include manufacturing engineers, computer programmers, and maintenance engineers.

Compared to the conventional manufacturing system, some benefits of FMS are listed in the following.

(1) Parts can be produced randomly, in batch sizes as small as one and at lower unit cost.

(2) Direct labor and inventories are reduced to yield major saving over conventional systems.

(3) The lead times required for product changes are shorter.

(4) Production is more reliable because the system is self-correcting and the product quality is uniform.

Section Ⅱ New Words and Phrases

document	*n.* 公文，文件，文档，档案，文献
routing	*n.* 路由选择；工艺路线
manually	*adv.* 手调，用手
labor-intensive	需大量劳动力的，劳动集约的
time-consuming	耗费时间的，旷日持久的
affix	*vt.* 使附于，粘贴
dedicated	*adj.* 专用的
monitor	*n.* 监听器，监视器，监控器，监测器
coordinate	*adj.* 同等的，并列的 *vt.* 协调
adjunct	*n.* 附件，助手 *adj.* 附属的
variant	*n.* 变量 *adj.* 变种的
database	*n.* 数据库，资料库
retrieve	*vt.* 找回，［计］检索，重新得到
generative	*adj.* 派生的，生产的，有生产能力的，再生的

comprehensive	*adj.* 全面的，广泛的，综合的
consistency	*n.* 一致性，连贯性
up-to-date	最新的，最近的，当代的，新式的
integrate…into	使……集成
subsystem	*n.* 次要系统，子系统
computer-integrated manufacturing	计算机集成制造
standardization	*n.* 标准化，格式化，规范化
reliability	*n.* 可靠性
legible	*adj.* 清晰的，易读的
incorporate	*vt.* 合并的，一体化的 *v.* 结合
CAPP（computer-aided process planning）	计算机辅助工艺设计
FMS（flexible manufacturing systems）	柔性制造系统
CNC（computer numerical control）	计算机数控
integrate	*v.* 结合；包含
material-handling	物料输送，原材料处理
successive	*adj.* 继承的，连续的
workstation	*n.* 工作站
optimize	*vt.* 使最优化
grinding	*n.* 磨削
cutting	*n.* 切削
powder metallurgy	粉末冶金
heat treating	热处理
finishing	*n.* 带式磨光，饰面，表面修饰，擦光
raw materials	原材料
inspection	*n.* 检查，检验，视察
assembly	*n.* 集合，装配，集会，集结，汇编
machine-tool	母机，机床
sophistication	*n.* 老练，成熟，精致，世故
configuration	*n.* 构造，结构，配置，外形
incorporate	合并的，结社的
yield	*n.* 产量，收益 *v.* 出产，生成
milling machine	铣床
drill press	钻床，立式钻床
grinder	*n.* 磨床，磨机，研磨机，磨工
punching	*n.* 凿孔，冲板，冲压，冲孔
shearing	*v.* 剪切
furnace	*n.* 炉子，熔炉

forging machine	锻造机
trimming press	冲拔罐修边机
retrieval system	回收系统
transfer mechanism	传输机械装置
pallet	*n.* 托盘，货盘
terminal	终点站，终端，接线端
idle	*adj.* 空闲的 *vi.*（机器）空转，闲置
schedule	*n.* 动态调度，工作调度
capital-intensive	资金集约型
debug	*vt.* 调试

Section Ⅲ Notes to Complex Sentences

[1] Process planning is concerned with the selecting methods of the production: tooling, fixtures, machinery, sequence of operations, and assembly.

工艺设计是有关生产方法的选择：工具、夹具、机器设备、操作顺序和装配。

be concerned with 意为“与某事有关，涉及某事物”。

[2] Computer-aided process planning accomplishes this complex task of process planning by viewing the total operation as an integrated system, so that the individual operations and steps involved in making each part are coordinated with others and are performed efficiently and reliably.

计算机辅助工艺设计是将全部工艺运行视为一个集成系统，在此基础上来完成工艺规划这个复杂的任务，这就使制造每一个零件时所涉及的单个操作和步骤能够与其他操作和步骤相协调，并且能被高效和可靠地完成。

involved in...是过去分词短语，修饰前面的名词 operations and steps。

[3] If the standard plan for a particular part is not in the computer files, a plan that is close to it, with a similar code number and an existing routing sheet, is retrieved.

如果计算机文档中没有某个特殊零件的标准工艺计划，则可调出一个与其相近的工艺计划，后者与前者具有相似编码，且存有工艺单。

[4] In the generative system, a process plan is automatically generated on the basis of the same logical procedures that would be followed by a traditional process planner in making that particular part.

在再生系统中，工艺计划会按照传统工艺员在制定特殊零件工艺时所遵循的相同逻辑程序而自动生成。

[5] Although currently it is used less commonly than the variant system, this system has such advantages as ① flexibility and consistency for process planning for new parts and ② higher overall planning quality, because of the capability of the decision logic in the system to optimize the planning and to utilize up-to-date manufacturing technology.

虽然目前它比其他系统应用得少，但是这个系统有以下优点：①适应性和一致性适合新零件的程序设计；②更高的总体设计品质，这是因为系统中的决策逻辑能优化工艺设计和利用最新的制造技术。

[6] These steps may involve one or more processes and operations (such as machining, grinding, cutting, forming, powder metallurgy, heat treating and finishing), as well as handling of raw materials, inspection and assembly. The most common applications of FMS to date have been in machining and assembly operations.

这些步骤可能包括一个或多个程序和操作（比如加工、磨削、切削、成形、粉末冶金、热处理和修整），还有处理原材料、检查和汇编。到目前为止，FMS 最常见的应用是机加工和装配操作。

as well as…意为“也，还”，常用来连接两个并列的成分，它强调的是前一项，后一项只是顺便提及。to date…意为“至今，到目前为止”。

[7] FMS can be regarded as a system which combines the benefits of two other systems: ① the highly productive but inflexible transfer lines and ② job-shop production, which can produce large product variety on stand-alone machines but inefficient.

柔性制造系统可以说是结合了另外两个系统的优点：① 高生产量但不灵活的输送线；② 能用独立机器生产大量多样化产品但效率低下的加工生产车间。

…which combines the benefits…是限定性定语从句，修饰 system。

which can produce large product…是非限定性定语从句，修饰 job-shop production。

[8] The system is capable of transporting the raw materials, blanks and parts in various stages of completion to any machine (in random order) and at any time.

这个系统能够在任何时间向任何机器（随机顺序）传输不同阶段加工完成的原材料、毛坯及零件。

[9] Dynamic scheduling is capable of responding to the quick changes in the product type and so is responsive to the real-time decisions.

动态调度能够对产品类型的快速变化做出反应，因此能对实时决策做出响应。

[10] However, the characteristics, performance and reliability of each unit in the system must be checked to ensure that parts moving from workstation to workstation are of the acceptable quality and dimensional accuracy.

此外，必须对系统中的每个制造单元的特点、性能和可靠性进行检验，确保在工作站之间流动的工件满足质量及尺寸精度方面的要求。

that parts moving from…是宾语从句，是 ensure 的宾语。

[11] This analysis should include such factors as the cost of capital, of energy, of materials and of labor, the expected markets for the products to be manufactured and any anticipated fluctuations in the market demand and product type.

这个分析应该包括这些因素，如资金、能源、材料和劳动力的成本、即将投产产品的市场预期以及在市场需求和产品类型方面可能出现的波动。

Section Ⅳ Exercises

Translate the following sentences into Chinese.

(1) When done manually, this task is highly labor-intensive and time-consuming and relies heavily on the experience of the process planner.

(2) The search for a standard plan is made in the database by a code number for the part; the plan is based on its shape and its manufacturing characteristics.

(3) Such a system is complex, however, because it must contain comprehensive and detailed knowledge of the part shape and dimensions, of the process capabilities, about selection of manufacturing methods, machinery, and tools, and of the sequence of operations to be performed.

(4) Flexible manufacturing systems represent the highest level of efficiency, sophistication and productivity that has been achieved in manufacturing plants.

(5) For machining operations, they usually consist of a variety of 3-5 axis machining centers, CNC lathes, milling machines, drill presses and grinders.

(6) Because FMS involves a major capital investment, efficient machine utilization is essential: machines must not stand idle. Consequently, the proper scheduling and process planning are crucial.

Section Ⅴ Supplementary Reading

Automated Process Planning

A practical process planning system with good decision rules for each activity will seldom generate bad plans. Such a practical system could be structured by reducing the decision-making process to a series of mechanical steps. However, even if a system of good decision rules were to be developed, the human interaction of man-variant process planning could still create problems. Process planning can become a boring and tedious job. Accordingly, man-variant planning often produces erroneous process plans. This, coupled with the labor intensity of man-variant planning, has led many industries to the investigation of the automation of the process planning. Whether a planning system is to be automated or manual, the seven general requirements of the process planning as defined by Doyle must be performed. In the following section, the framework for the automated process planning will be described.

Spur and Optiz were among the first to write on the automation of the manufacturing systems and the role that process planning should play in these systems. Spur was perhaps the first to define variant and generative methods of the process planning and the mechanization and implementation of such planning systems. The variant method of the process planning is based on the principle of group technology and essentially consists of the following two steps.

(1) Build a catalog (or "menu" as it is often called) of the process plans to produce a gamut of parts, given a set of machine tools.

(2) Create the software necessary to examine the part that is being planned and find the closest facsimile in the catalogue, then retrieve the associated process plans.

The generative method of the process planning essentially consists of the following four steps.

(1) Describe a part in detail.

(2) Describe a catalog of process available to produce the parts.

(3) Describe the machine tool(s) that can perform these processes.

(4) Create the software to inspect the part, process, and available machinery to determine whether all three are compatible.

In general, planning using generative principles requires a detailed description of the part as well as a detailed understanding of manufacturing process and their accuracy. Manufacturing plans based on the variant principle are determined by activating several standard solutions for individual operations and adapting or adjusting them where necessary.

A flexible manufacturing system is a manufacturing system in which there is some amount of the flexibility that allows the system to react in the case of changes, whether predicted or unpredicted. This flexibility is generally considered to fall into two categories, which both contain the numerous subcategories.

The first category, machine flexibility covers the systems ability to be changed to produce the new product types and ability to change the order of the operations executed on a part. The second category is called the routing flexibility, which consists of the ability to use multiple machines to perform the same operation on a part, as well as the systems ability to absorb large-scale changes, such as in volume, capacity or capability.

Most FMS systems comprise of three main systems. The work machines which are automated by the CNC machines are connected by a material handling system to optimize the parts flow and the central control computer which controls the material movements and machine flow.

The main advantage of an FMS is its high flexibility in managing manufacturing resources like time and effort in order to manufacture a new product. The best application of an FMS is found in the production of small sets of products like those from a mass production.

As a summary of the above, the advantages and disadvantages of FMS are listed below.

(1) Preparation time for new products is shorter due to flexibility.

(2) Improved production quality.

(3) Saved labour cost.

(4) Productivity increment.

(5) Saved labour costs must be weighed against the initial cost of FMS.

(6) Drawback of increased flexibility may be decreased productivity.

Exercises

1. Write T(True) or F(False) beside the following statements about the text.

(1) ____________ A flexible manufacturing system integrates all major elements of the manufacturing to a highly automated system.

(2) ____________ Different computer instructions for the manufacturing process can't be downloaded for all the successive parts passing through the workstation.

(3) __________The most common applications of FMS to date have been in the manufacturing and production operations.

(4) ___________Flexible manufacturing systems represent the lowest level of efficiency, sophistication and productivity that has been achieved in the power plants.

(5) __________The types of machines in workstations depend on the type of production. For machining operations, they usually consist of a variety of 3−5 axis machining centers, CNC lathes, milling machines, drill processes and grinders.

(6) __________Material handling isn't controlled by a central computer and not performed by automated guided vehicles, conveyors, and various transfer mechanisms.

(7) __________This sub-system controls the machinery and equipment in workstations and the transporting of raw materials, blanks and parts in various stages of completion from machine to machine.

(8) __________However, the characteristics, performance and reliability of each unit in the system mustn't be checked to ensure that parts moving from workstation to workstation aren't of the acceptable quality and dimensional accuracy.

2. Choose the best answer.

(1) Different computer__________for the manufacturing process can be downloaded for each successive part passing through the workstation.

A. instructions　　B. directions　　C. orders　　D. commands

(2) The most common___________of FMS to date have been in machining and assembly operations.

A. use　　B. adoption　　C. employment　　D. application

(3) The__________of FMS is such that it can handle a variety of part configurations and produce them in any order.

A. freedom　　B. elasticity　　C. capability　　D. flexibility

(4) This sub-system controls the machinery and equipment in workstations and the__________ of raw materials, blanks and parts in various stages of completion from machine to machine.

A. delivery B. transportation C. sending D. uploading

(5) Such a system is complex, however, because it must contain comprehensive and detailed knowledge of the part shape and dimensions, of the process capabilities, about selection of__________, machinery, and tools, and of the sequence of operations to be performed.

A. manufacturing ways B. manufacturing philosophy

C. manufacturing management D. manufacturing approaches

3. Fill in the blanks with words or expressions according to the text.

(1) __________represent the highest level of efficiency, sophistication and productivity that has been achieved in __________.

(2) For __________, they usually consist of a variety of__________machining centers, CNC lathes, milling machines, drill presses and grinders.

(3) However, even if a system of good decision rules were to be developed, the human __________ of __________could still create problems.

(4) Spur was perhaps the first to define__________ methods of the process planning and the __________of such planning systems.

(5) A flexible manufacturing system is a__________in which there is some amount of the__________that allows the system to react in the case of changes, whether predicted or unpredicted.

(6) The work machines which are automated by the CNC machines are connected by a __________ to optimize the parts flow and the central control computer which controls the material movements and __________.

4. Answer the following questions according to the text.

(1) 什么是柔性制造系统？柔性制造系统的生产过程是什么？柔性生产系统的优点是什么？

(2) 柔性制造系统的基本元素有什么？柔性制造系统为什么很重要？

(3) 柔性制造系统如何安排调度？柔性制造系统的经济合理性如何体现？

5. Translate the following paragraphs into Chinese.

(1) Other types of operations suitable for FMS include sheet metal forming, punching and shearing and forging; they incorporate furnaces, forging machines, trimming presses, heat-treating facilities and cleaning equipment.

(2) In general, planning using generative principles requires a detailed description of the part as well as a detailed understanding of manufacturing process and their accuracy. Manufacturing plans based on the variant principle are determined by activating several standard solutions for individual operations and adapting or adjusting them where necessary.

(3) The first category, machine flexibility covers the systems ability to be changed to produce the new product types and ability to change the order of the operations executed on a part. The second category is called the routing flexibility, which consists of the ability to use multiple

machines to perform the same operation on a part, as well as the systems ability to absorb large-scale changes, such as in volume, capacity or capability.

单元评价

通过本单元的学习，使学生掌握机械制造工艺及仿真相关的英语词汇及常用英语表达语句，并具备运用英语检索机械制造工艺及仿真相关信息及使用英语描述机械制造工艺及仿真相关内容的能力，同时培养学生制造强国、科技强国的使命担当意识。

单元 8

数控机床及应用

【学习目标】

知识目标：

1. 熟悉数控机床及应用的相关英语词汇。

2. 掌握与人交流数控机床及应用时常用的英语表达。

技能目标：

1. 具备数控机床及应用英语词汇寻读和跟读技巧，能够运用英语检索相关信息。

2. 具备使用英语描述数控机床及应用相关信息和过程的能力。

素养目标：

1. 培养学生的质量意识、安全意识。

2. 培养学生的沟通能力及团队协作精神。

3. 培养学生崇德向善、诚实守信、爱岗敬业、精益求精、吃苦耐劳、无私奉献的工匠精神。

Section Ⅰ Texts

One of the most important developments in the manufacturing automation is the numerical control(NC). NC has been defined by the Electronics Industries Association (EIA) as "a system in which the actions are controlled by the direct insertion of the numerical data at some point. The system must automatically interpret at least some the portion of this data." The data required to produce a part is called a part program. The part program is a group of instructions read by the control system and is converted into the signals that move the drives.[1] The focus of the NC machines has traditionally been on the complex parts manufactured in large volumes. However, because of the development of the more efficient programming languages, NC is now employed for smaller volume sizes. A numerical-control machine tool system includes the machine-control unit (MCU). The MCU is further divided into two elements, the data-processing unit (DPU) and the control-loops unit (CLU). The DPU processes the coded data read from the tape or other data-store media and passes information on the position of each axis, its direction of motion, feed and auxiliary function controls signals to the CLU.[2] The CLU operates the actual position and velocity of each of the axes and indicates an operation's completion time. The DPU sequentially reads the data when each line has completed execution as noted by the CPU.

The motion control of the NC machine tools is completed by translating NC codes into machine commands. The NC codes can be broadly classified into two groups: ①commands for controlling the individual machine components, such as motor on/off control (these tasks are accomplished by sending electric pulses to the relay system or logic control network) and ②commands for controlling the relative movement of the workpiece and the tools. These commands consist of the information, such as axis and distance to be moved at each specific time unit. They are translated into the machine-executable motion-control commands that are then carried out by the electro-mechanical control system.[3]

Currently, NC controllers are all built with the use of computer technology. Such controllers are called computer numerical control(CNC, as shown in Fig. 8-1-1). CNC systems are more flexible than their NC counterparts because they allow programs to be edited, stored in the memory and recalled instantly. CNC machines can generally machine more complex shapes than NC machines can and also provide circular interpolation. CNC controllers have been applied to nearly every kind of machine tools, lathes, milling machines, drill presses, grinders, etc. Features available to the modern machines include tools, pallet, and workpiece changers. Controller features include interpolators, graphics interfaces, interactive operator programming(Fig. 8-1-2) and data communication.

Fig. 8-1-1 CNC

Fig. 8-1-2 CNC programming

In numerical control, data concerning all aspects of the machining operation, such as locations, speeds, feeds and cutting fluids, are stored in compact disks, floppy or computer's hard disks. The concept of NC is that the specific information can be relayed from these storage devices to the machine tool's control unit. On the basis of the information input, the relays and other devices (called hard-wired controls) can be actuated to obtain a desired machine setup. Complex operations, such as turning a part having various contours or die sinking in a milling machine, can be carried out. If a single computer is used to download the part programs to a variety of NC machines, this system is referred to as a direct numerical-control (DNC) system.[4] This computer may also be used to download instructions to the material-handling system equipment.

A number of specific functions are executed by the components of an NC-controlled servo drive system. An operating panel/keyboard allows the alpha-numeric instructions to be entered in the machine. A decoder/encoder receives the data, entered from the computer and divides them into two sections: one for the part geometry data and the other for the process data which includes information about feed rates, spindle speeds and other machining parameters.[5] The geometric data also contains information about the tool motions. The same set of data is also used to determine the tool-length, tool-radius, tool-compensation, etc, required in the process. Process data consists of switching functions for adjusting feed rates, spindle speeds, tool changes, cutting fluid application, etc. The switching functions are initiated by the switching commands to an interface unit, where they are compared with the feedback signals from the machine tool and then are translated into the appropriate control signals for the particular device to be operated.[6] In addition, a linkage is provided with the safety switches as a hazard precaution, in order to stop the machine in case of the conflicting instructions, which may damage the machine or cause injury. The geometric data can be used only after adjustments are made in order to fit the particular tool-workpiece relationship. This "fitting" allows the programmer to edit a program irrespective of the actual position of the workpiece, etc. Corrective calculations, such as the length of the drill, the size of the turning tool or the diameter of the milling cutters, etc. , could be made regarding the dimensions of the tool.

Section Ⅱ New Words and Phrases

numerical control (NC)	数字控制
Electronics Industries Association (EIA)	电子工业协会
machine-control unit (MCU)	机床控制单元
data-processing unit (DPU)	数据处理单元
control-loops unit (CLU)	操纵系统单元
computer numerical control (CNC)	计算机数字控制
direct numerical-control (DNC)	直接数字控制
insertion	*n.* 插入
interpret	*v.* 解释，说明
convert into	使转变为
volume	*n.* 卷，册，体积，量，大量，音量
coded	*adj.* 电码的，编码的
auxiliary	*adj.* 辅助的，补助的
velocity	*n.* 速度，速率，迅速，周转率
flexible	*adj.* 柔韧的，灵活的，柔软的，可通融的
pallet	*n.* 扁平工具，棘爪，货盘

interpolator	*n.* 插值，插入；［计］校对机，分类机
floppy	*adj.* 软的
floppy disk	软盘
contour	*n.* 轮廓，周线，等高线
material-handling	物料输送，材料处理
servo	*n.* 伺服，伺服系统
alpha-numeric instruction	字符指令
decoder	*n.* 解码器
encoder	*n.* 译码器，编码器
feed rate	馈送率
spindle speed	转轴速度
parameter	*n.* 参数，参量
feedback	*n.* 反馈，反应
linkage	*n.* 连接
interface unit	接口部件，接口装置，连接器件

Section Ⅲ Notes to Complex Sentences

[1] The part program is a group of instructions, read by the control system, and is converted into the signals that move the drives.

零件加工程序是一组指令，由控制系统读取，然后转变成运行驱动装置的信号。

[2] The DPU processes the coded data read from the tape or other data-store media and passes the information of the position of each axis, its direction of motion, feed and auxiliary function controls signals to the CLU.

DPU（数据处理单元）处理从光盘或其他数据储存介质读取的数据编码，然后将有关每个轴的位置、运动方向、进给量和辅助功能控制信号等信息传递给 CLU。

read from…data-store media 为 coded data 的后置定语。

passes information to…：将信息传递给……。

[3] They are translated into the machine-executable motion-control commands that are then carried out by the electro-mechanical control system.

它们被翻译成由机床执行的运动控制指令，然后由机电控制系统执行这个指令。

that 引导的定语从句说明 commands。

carry out 意为“实行，执行，实现”。

E.g., no matter what difficulty you will meet with, carry out your plan.

不管你遇到什么困难，都要执行计划。

[4] If a single computer is used to download the part programs to a variety of NC machines, this system is referred to as a direct numerical-control (DNC) system.

如果用一台单独计算机下载零件加工程序到多台 NC 机床，则这个系统被称为直接数字控制系统。

[5] A decoder/encoder receives the data entered from the computer and divides them into two sections: one for the part geometry data and the other for the process data which includes information about feed rates, spindle speeds and other machining parameters.

一台译码器/编码器接收到从计算机输入的数据，然后将其分成两部分：一部分是零件的几何尺寸数据；另一部分是工艺数据，包括进给率、转轴速度和其他加工参数。

句中，which 引导的定语从句说明 data。

[6] The switching functions are initiated by switching commands to an interface unit, where they are compared with the feedback signals from the machine tool and then are translated into the appropriate control signals for the particular device to be operated.

开关功能由开关指令启动接口装置，与从机床反馈的信号比较，然后翻译成适当的控制信号以操作特殊装置。

句中，where 引导的定语从句用来说明 interface unit。

Section Ⅳ Exercises

Translate the following sentences into Chinese.

(1) The CLU operates the actual position and velocity of each of the axes and indicates an operation's completion time.

(2) In numerical control, data concerning all aspects of the machining operation, such as locations, speeds, feeds, and cutting fluids, are stored in compact disks, floppy or computer's hard disks.

(3) In addition, a linkage is provided with safety switches as a hazard precaution, in order to stop the machine in case of the conflicting instructions, which may damage the machine or cause injury.

Section Ⅴ Supplementary Reading

Operational Sequence

NC starts with the parts programmer who, after studying the engineering drawing, visualizes the operations needed to machine the workpiece. These instructions, commonly called a program, are prepared before the part is manufactured and consist of a sequence of symbolic codes that specify the desired action of the tool workpiece and machine. Even in computer aided design and manufacturing, this interpretation is necessary. The engineering drawing of the workpiece is examined and the processes are selected to transform the raw material into a finished part that meets the dimensions, tolerances, and specifications. This process planning is concerned with the preparation of an operations sheet or a route sheet or traveler. These different titles describe the procedure or the sequence of the operations and it lists the machines, tools, and operational costs. The particular order is important. Once the operations are known, those that pertain to NC are further engineered in that detail sequences are selected.

A program is prepared by listing codes that define the sequence. A part programmer is trained about manufacturing process and is knowledgeable of the steps required to machine a part and documents these steps in a special format. These are two ways to program for NC, by manual or computer assisted part programming. The part programmer must understand the processor language used by the computer and the NC machine.

If manual programming is required, the machining instructions are listed in a form called a part program manuscript. This manuscript gives instructions for the cutter and workpiece, which must be positioned relative to each other for the path instructions to machine the design. Computer assisted part programming, on the other hand, does much of the calculation and translates brief instructions into a detailed instruction and coded language for the control tape. Complex geometries, many common hole centers and symmetry of the surface treatment can be simply programmed under the computer assistance, which saves programmer time.

Tape preparation is next, as the program is "typed" onto a tape or punched card. If the programming is manual, the 1 in (25 mm) wide perforated tape is prepared from the part manuscript on a typewriter with a standard keyboard but equipped with a punch device capable of punching holes along the length of the tape. If the computer is used, the internal memory interprets the programming steps, does the calculations to provide a listing of the NC steps and additionally will prepare the tape. Some tapes contain electronic or magnetic signals and other systems use disks or direct computer inputs.

Verification is the next step, as the tape is run through a computer and a plotter will simulate the movements of the tool and graphically display the final paper part which is often in a two-dimensional layout describing the final part dimensions. This verification uncovers major mistakes.

The final step is production using the NC tape. This involves ordering special tooling, fixtures and scheduling the job. A machine operator loads the tape onto a program reader that is part of the machine-control unit often called an MCU. This converts coded instructions into machine tool actions. The media that is loaded onto the MCU can be perforated tape, magnetic tape, tabulating cards, floppy disks, or direct computer signals from other computers or satellites. Perforated paper tape is the predominant input medium, but the concepts are the same whatever the input.

Exercises

1. Write T(True) or F(False) beside the following statements about the text.

(1) ________One of the most important developments in the manufacturing automation is

the numerical control(NC).

(2) __________The part program is a group of instructions read by the control system and is converted into the signals that move the drives.

(3) __________The DPU processes the coded identification read from the tape or other database and transports information on the direction of the axis, its direction of motion, feed and auxiliary function controls signals to the CLU.

(4) __________They are changed into the machine-executable motion-control instructions that are then carried out by the electro-mechanical control devices.

(5) __________ Controller features include interpolators, graphics interfaces, interactive operator programming and data communication.

(6) __________The concept of NC is that the comprehensive information can be shifted from these databases to the machine tool's control unit.

(7) __________The switching functions are initiated by the switching commands to an interface unit, where they are compared with the feedback signals from the machine tool and then are translated into the appropriate control signals for the particular device to be operated.

(8) __________Corrective calculations, such as the depth of the drill, the size of the turning tool or the diameter of the milling cutters, etc. , must be made without the regulations of the tool.

2. Choose the best answer.

(1) The part program is a group of__________read by the control system and is converted into the signals that move the drives.

A. instructions B. directions C. orders D. commands

(2) However, because of the development of the more efficient programming languages, NC is now employed for smaller __________.

A. volume amount B. volume scale C. volume sizes D. volume accumulation

(3) The concept of NC is that the specific information can be relayed from these __________ to the machine tool's control unit.

A. storage equipment B. storage apparatus C. storage database D. storage carrier

(4) The geometric__________can be used only after adjustments are made in order to fit the particular tool-workpiece relationship.

A. statistics B. information C. presentation D. data

(5) In addition, a linkage is provided with safety switches as a__________, in order to stop the machine in case of the conflicting instructions, which may damage the machine or cause injury.

A. hazard prevention B. hazard precaution C. hazard warning D. hazard information

3. Fill in the blanks with words or expressions according to the text.

(1) NC starts with the__________who, after studying the engineering drawing, visualizes the operations needed to machine the__________.

(2) This __________ is concerned with the preparation of an __________or a route sheet or

traveler.

(3) A __________ is trained about __________ and is knowledgeable of the steps required to machine a part and documents these steps in a special format.

(4) If __________ is required, the machining instructions are listed in a form called a part __________.

(5) Complex geometries, many common hole centers and symmetry of the __________ can be simply programmed under the __________, which saves programmer time.

(6) __________ is the predominant __________, but the concepts are the same whatever the input.

4. Answer the following questions according to the text.

(1) 什么是数字化控制？数控机床的运动原理是什么？

(2) CNC 系统柔性程度为什么高？它可以应用在哪些领域？

(3) 数控的概念是什么？NC 系统控制的伺服驱动系统是如何工作的？

5. Translate the following paragraphs into Chinese.

(1) In numerical control, data concerning all aspects of the machining operation, such as locations, speeds, feeds, and cutting fluids, are stored in compact disks, floppy or computer's hard disks.

(2) Verification is the next step, as the tape is run through a computer and a plotter will simulate the movements of the tool and graphically display the final paper part which is often in a two-dimensional layout describing the final part dimensions. This verification uncovers major mistakes.

(3) A program is prepared by listing codes that define the sequence. A part programmer is trained about manufacturing process and is knowledgeable of the steps required to machine a part and documents these steps in a special format. These are two ways to program for NC, by manual or computer assisted part programming. The part programmer must understand the processor language used by the computer and the NC machine.

单元评价

通过本单元的学习，使学生掌握数控机床及应用相关的英语词汇及常用表达语句，并具备运用英语检索数控机床及应用相关信息及使用英语描述数控机床及应用相关内容的能力，同时培养学生崇德向善、诚实守信、爱岗敬业、精益求精、吃苦耐劳、无私奉献的工匠精神。

单元 9

PLC 控制技术与应用

【学习目标】

知识目标：

1. 熟悉 PLC 与控制系统相关英语词汇。

2. 掌握与技术人员交流工业现场控制、可编程控制器时常用的英语表达语句。

技能目标：

1. 具备可编程逻辑控制器相关英语词汇跟读技巧，能够运用英语检索控制系统及 PLC 相关学术资料。

2. 具备使用英语描述 PLC 与工业现场控制的能力。

素养目标：

1. 遵守职业道德准则和行为规范。

2. 树立爱岗敬业、全力以赴的职业道德风尚。

3. 提高学生交流沟通、团队协作的职业精神。

Section Ⅰ Texts

1. What Is PLC

A programmable logic controller (PLC) (Fig. 9-1-1) is a device that was invented to replace the necessary sequential relay circuits for the machine control. The PLC works by looking at its inputs and depending upon their state, turning on/off its outputs.[1] The user enters a program, usually via software, which gives the desired results.[2]

PLCs are used in many "real world" applications. If there is industry present, chances are good that there is a PLC present. If you are involved in machining, packaging, material handling, automated assembly or countless other industries, you are probably already using them. If you are not, you are wasting money and time. Almost any application that needs some type of electrical control has a need for a PLC.

For example, let's assume that when a switch turns on we want to turn a solenoid on for 5s and then turn it off regardless of how long the switch is on. We can do this with a simple external timer.

But what if the process includes 10 switches and solenoids? We would need 10 external timers. What if the process also needed to count how many times the switches individually turned on?[3] We need a lot of external counters. As you can see, the bigger the process is, the more of a need we have for a PLC. We can simply program the PLC to count its inputs and turn the solenoids on for the specified time.

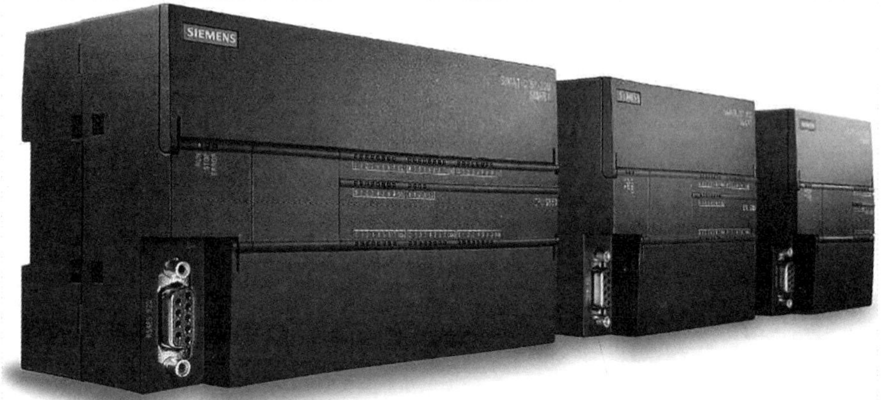

Fig. 9-1-1 Programmable logic controller

2. PLC History

In the late 1960s, PLCs were first introduced. The primary reason for designing such a device was eliminating the large cost involved in replacing the complicated relay based machine control systems. Bedford Associates (Bedford, MA) proposed something called a modular digital controller (MODICON) to a major or U. S. car manufacturer. Other companies at the time proposed computer based schemes, one of which was based upon the PDP-8. The Modicon 084 brought the world's first PLC into the commercial production.

When the production requirements changed, so did the control system.[4] This becomes very expensive when the change is frequent. Since the relays are mechanical devices, they also have a limited lifetime which required the strict adhesion to the maintenance schedules. Troubleshooting was also quite tedious when so many relays are involved. Now, picture a machine control panel that included many, possibly hundreds or thousands of individual relays.[5] The size could be mind boggling. How about the complicated initial wiring of so many individual devices! These relays would be individually wired together in a manner that would yield the desired outcome. Were there problems? You bet!

These "new controllers" also had to be easily programmed by the maintenance and plant engineers. The lifetime had to be long and the programming changes should easily be performed. They also had to survive in the harsh industrial environment. That's a lot to ask! The answers were to use a programming technique most people were already familiar with and replaced the mechanical parts with the solid-state ones.

In the mid 1970s, the dominant PLC technologies were the sequencer state-machines and the bit-slice based CPU. The AMD 2901 and 2903 were quite popular in Modicon and AB PLCs.

Communications abilities began to appear in, approximately, 1973. The first such system was Modicon's Modbus. The PLC could now talk to other PLCs and they could be far away from the actual machine they were controlling. They could also now be used to send and receive varying voltages to allow them to enter the analog world. Unfortunately, the lack of standardization coupled with continually changing technology has made PLC communications a nightmare of incompatible protocols and physical networks. Still, it was a great decade for the PLC! Fig. 9-1-2 shows the communication protocol of PLC.

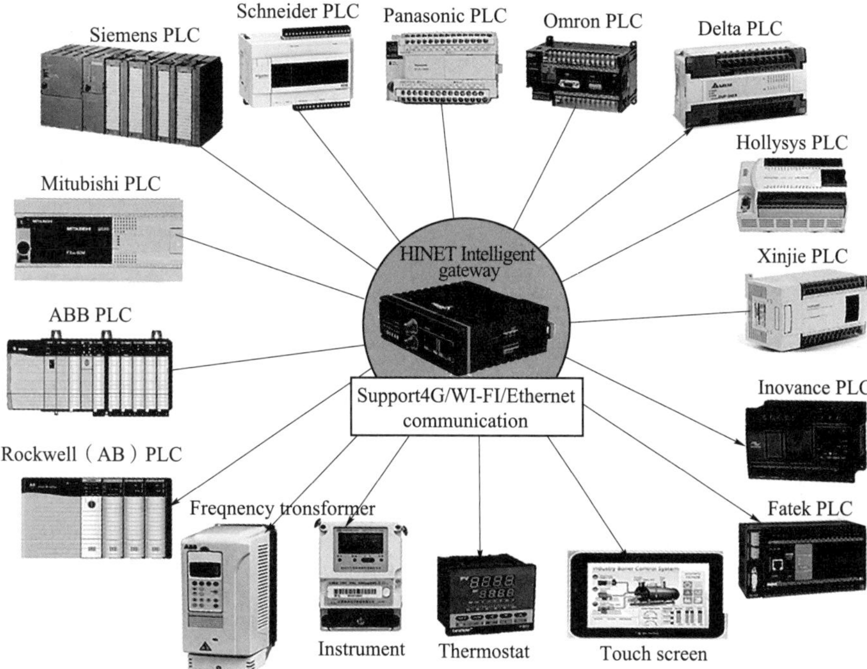

Fig. 9-1-2 Communication protocol of PLC

The 1980s saw an attempt to standardize communications with General Motor's manufacturing automation protocol(MAP).[6] It was also a time for reducing the size of the PLC and making their software programmable through symbolic programming on personal computers (PCs) instead of dedicated programming terminals or handheld programmers. Today, the world's smallest PLC is about the size of a single control relay!

The 1990s have seen a gradual reduction in the introduction of new protocols and the modernization of the physical layers of some of the more popular protocols that survived the1980s. The latest standard (IEC 1131-3) has tried to merge PLC programming languages under one international standard. We now have PLCs that are programmable in function block diagrams, instruction lists, C and structured text all at the same time! PCs are also being used to replace PLCs in some applications. The original company who commissioned the Modicon 084 has actually switched to a PC based control system.

What will the next decade bring? Only time will tell.

Section Ⅱ New Words and Phrases

sequential	*adj.* 顺序的
relay	*n.* 继电器
assume	*v.* 假设
solenoid	*n.* 电磁铁，螺线管
regardless	*adv.* 不管
timer	*n.* 定时器
counter	*n.* 计数器
propose	*v.* 提议，发起
scheme	*n.* 方案
frequent	*adj.* 频繁的
adhesion	*n.* 黏附，与……保持一致
maintenance	*n.* 维护
schedule	*n.* 计划表，日程
troubleshoot	*v.* 故障排除，检修
tedious	*adj.* 乏味的
picture	*v.* 勾勒，想象
panel	*n.* 面板
boggle	*vi.* 吃惊，损坏
mind boggling	令人震惊
yield	*v.* 生成
outcome	*n.* 结果
bet	*v.* 打赌
harsh	*adj.* 糟糕的，困难的，恶劣的
solid-state	*adj.* 固态的
slice	*n.* 一片，一条
couple	*v.* 耦合，成对
nightmare	*n.* 噩梦
incompatible	*adj.* 不兼容的
protocol	*n.* 协议
symbolic	*adj.* 符号的，象征性的
dedicated programming terminals	专用编程终端
gradual	*adj.* 逐渐的

layer	*n.* 层
merge	*v.* 并入
block diagram	框图
commission	*n.* & *v.* 订做，专做

Section Ⅲ Notes to Complex Sentences

[1] The PLC works by looking at its inputs and depending upon their state, turning on/off its outputs.

PLC 通过检测其输入并根据输入的状态，接通或断开其输出。

work 为谓语，by...意为“通过……”，分词短语 turning...作 by 的宾语。全句相当于“The PLC works by looking at its inputs and turning on/off its outputs depending upon their state”。

[2] The user enters a program, usually via software, which gives the desired results.

用户输入一个程序，通常是通过软件进行，该程序给出期望的结果。

enter：键入；via：经由，通过；which 引出定语从句，说明 program。

[3] What if the process also needed to count how many times the switches individually turned on?

如果该过程还需要计算开关分别接通的次数呢？

What if...？如果……怎么办？how many times 作 count 的宾语。

[4] When production requirements changed, so did the control system.

当生产需求变更时，控制系统也要随之变更。

so did...是由 so 引导的倒装语句，谓语 did 代替 changed.

[5] Now, picture a machine control panel that include many, possibly hundreds or thousands of the individual relays.

现在想象一台机器的控制面板，它含有几百个或者几千个继电器。

picture：想象。本句无主语，是祈使句结构。

[6] The 1980s saw an attempt to standardize the communications with the General Motor's manufacturing automation protocol(MAP).

20 世纪 80 年代出现了以通用汽车公司的生产自动化协议（MAP）为通信标准化的尝试。

Section Ⅳ Exercises

Translate the following paragraph into Chinese.

A PLC's control loop is a continuous cycle of the reading inputs, solving the ladder logic and then changing the outputs. Like any other computer, this does not happen instantly. When the power is turned on initially, the PLC does a quick sanity check to ensure that the hardware is working properly. If there is a problem, the PLC will halt and indicate there is an error. For example, if the PLC backup battery is low and the power is lost, the memory will be corrupt and this will result in a fault. If the PLC passes the sanity check, it will then scan all the inputs. After the input values are stored in memory, the ladder logic will be scanned. When it is complete, the outputs will be scanned and the output values will be refreshed. After this, the system goes back to the first step (doing a sanity check) and the loop goes on indefinitely.

Section V Supplementary Reading

What Is a PLC?

The PLC or programmable logic controller is a computer with a single mission. Most commonly used in industrial applications, it usually lacks a monitor, keyboard and a mouse, as it is normally programmed to operate a machine or system using but one program. Factory assembly line machinery is activated and monitored by a single PLC, where in the past hundreds of timers and relays would have been required to do the task. The machine or system user rarely, if ever, interacts directly with the PLC's program. When it is necessary to either edit or create the PLC program, a personal computer is usually (but not always) connected to it.

In the 1960s, the American automotive industry was searching for a way to do business better. The processes of sequencing, control and interlock logic needed for the automobile manufacturing was a time consuming and arduous task, which required the manual updating of the relays, timers and dedicated closed-loop controllers. When a new model was coming off the drawing board, skilled electricians were called on to reset the production line. GM Hydramatic specifically requested proposals for a replacement for the old system that would speed up the process and keep costs down.

Bedford Associates, out of Bedford, Massachusetts, came up with the winning proposal, designated the 084. The 084 became the first PLC and a new industry was born.

Programmable logic controllers, sometimes be referred to simply as the programmable controllers, are microprocessor based units that, in essence, monitor external sensory activity from additional devices. They take in the date, which reports on a wide variety of activity such as machine performance, energy output, and process impediment. They also control the attached motor starters, pilot lights, valves and many other devices. Both functions respond to a custom, user-created program.

Programmable logic controllers contain a variable number of input/output (I/O) ports, and are typically reduced instruction set computer (RISC) based. They are designed for real-time use, and often must withstand harsh environments on the shop floor. The programmable logic controller circuitry monitors the status of multiple sensor inputs, which control the output actuators, which may be things like motor starters, solenoids, lights and displays or valves.

Most PLCs are programmed in a special language called ladder logic. Ladder logic is essentially a Boolean logic-solving program with a graphical user interface designed to look like an electrician. A modern programmable logic controller is usually programmed in any one of several languages, ranging from ladder logic to Basic or C. Typically, the program is written in a development environment on a personal computer, and then is downloaded onto the programmable logic controller directly through a cable connection. The program is stored in the programmable logic controller in non-volatile memory.

Exercises

1. Write T(True) or F(False) beside the following statements about the text.

(1) ________ A PLC (i. e., programmable logic controller) is a device that has been invented to replace the unnecessary sequential relay currents for the machine control.

(2) ________ If you are involved in machining, packaging, material handling, automated assembly or countless other industries you are probably already using them.

(3) ________ The primary reason for designing such a device was eliminating the small cost involved in replacing the complicated relay based carrying devices and the control systems.

(4) ________ These relays would be identically wired together in a manner that would abandon the unpredictable outcome.

(5) ________ They could also now be used to send and receive varying voltages to allow them to enter the analog world.

(6) ________ The concept of NC is that the comprehensive information can be shifted from these databases to the machine tool's control unit.

(7) ________ It was also a time for increasing the size of the PLC and making them hardware programmable through typical programming on most of the computers instead of the tailored programming terminals or handheld people.

(8) ________ When it is complete, the outputs will be presented and the input values will be updated.

2. Choose the best answer.

(1) Since the relays are ________, they also have a limited lifetime which required the strict adhesion to the maintenance schedules.

A. mechanical approaches B. mechanical equipment

C. mechanical devices D. mechanical container

(2) The answers were to use a programming technique most people were already familiar with and replaced the mechanical parts with the ________ ones.

A. solid-state B. solid-induced

C. solid-reduced D. solid-started

(3) Unfortunately, the lack of__________coupled with continually changing technology has made PLC communications a nightmare of incompatible protocols and physical networks.

A. standard B. standardization C. standardly D. standary

(4) Most commonly used in__________, it usually lacks a monitor, keyboard and a mouse, as it is normally programmed to operate a machine or system using but one program.

A. industrial use B. industrial purpose

C. industrial destination D. industrial applications

(5) The programmable logic controller circuitry monitors the status of multiple sensor inputs, which control the output__________, which may be things like motor starters, solenoids, lights and displays or valves.

A. actuators B. actuation C. actual D. actually

3. Fill in the blanks with words or expressions according to the text.

(1) If you are involved in machining, packaging, material handling, automated assembly or __________ other __________ you are probably already using them.

(2) The__________for designing such a device was eliminating the large cost involved in replacing the complicated relay based __________.

(3) These relays would be individually wired together in a manner that would__________ the desired __________.

(4) If__________is required, the machining instructions are listed in a form called a part __________.

(5) The processes of sequencing, control and interlock logic needed for the __________ was a __________task, which required the manual updating of the relays, timers and dedicated closed-loop controllers.

(6) Typically, the program is written in a__________ on a personal computer, and then is downloaded onto the programmable logic controller directly through a__________.

4. Answer the following questions according to the text.

(1) PLC 是什么？PLC 在现实中有哪些应用？

(2) PLC 是如何实现通信的？可编控制器的工作原理是什么？

(3) 梯形图是什么？它与 Basic 和 C 语言有什么关系？

5. Translate the following paragraphs into Chinese.

(1) These "new controllers" also had to be easily programmed by the maintenance and plant engineers. The lifetime had to be long and the programming changes should easily be performed. They also had to survive the harsh industrial environment. That's a lot to ask! The answers were to use a programming technique most people were already familiar with and replaced the mechanical parts with the solid-state ones.

(2) The PLC or programmable logic controller is a computer with a single mission. Most

commonly used in industrial applications, it usually lacks a monitor, keyboard and a mouse, as it is normally programmed to operate a machine or system using but one program. Factory assembly line machinery is activated and monitored by a single PLC, where in the past hundreds of timers and relays would have been required to do the task. The machine or system user rarely, if ever, interacts directly with the PLC's program. When it is necessary to either edit or create the PLC program, a personal computer is usually (but not always) connected to it.

(3) Most PLCs are programmed in a special language called ladder logic. Ladder logic is essentially a Boolean logic-solving program with a graphical user interface designed to look like an electricians. A modern programmable logic controller is usually programmed in any one of several languages, ranging from ladder logic to Basic or C. Typically, the program is written in a development environment on a personal computer, and then is downloaded onto the programmable logic controller directly through a cable connection. The program is stored in the programmable logic controller in non-volatile memory.

单元评价

通过本单元课文阅读，单词、短语练习，扩展阅读及习题练习，使学生掌握与 PLC 与工业现场控制相关的词汇和表达语句，能够自行完成相关文献内容的翻译，并具备与技术人员进行英语基础交流的能力，同时培养学生交流沟通、团队协作的职业精神。

单元 10

工业机器人及应用

【学习目标】

知识目标：

1. 熟悉工业机器人及其应用领域的相关英语词汇。

2. 掌握与人交流工业机器人时常用的英语表达语句。

技能目标：

1. 具备工业机器人及其应用领域相关英语词汇跟读技巧，能够运用英语检索工业机器人相关研究文献。

2. 具备使用英语描述工业机器人及其应用领域的能力。

素养目标：

1. 具备搜集工业机器人相关技术资料，尽快熟悉新设备相关知识的能力。

2. 具备独立学习、灵活运用所学知识分析并解决问题的能力。

Section I Texts

1. Definition

"A robot is a reprogrammable, multifunctional machine designed to manipulate materials, parts, tools or specialized devices through the variable programmed motions for the performance of a variety of tasks."

— Robotics Industries Association

"A robot is an automatic device that performs functions normally ascribed to humans or a machine in the form of a human."

— Webster's Dictionary

2. History

The word"robot"was coined from the Czech word for forced labor or serf. The term"robotics" refers to the study and use of robots.

The first industrial modern robots were the Unimates developed by George Devol and Joe Engelberger in the late 1950's and early 1960's. Engelberger founded Unimation and was the first to market the robots. As a result, he has been called the "father of robotics".

3. Key Components

Although the robots differ in a wide variety, they are composed in the same way in the infrastructure. The key components of a robot are listed below.

(1) Power conversion unit—provide power to any other parts of a robot.

(2) Sensors—measure robot configuration/condition and its environment and send such information to robot controller as electronic signals (e. g., arm position, presence of toxic gas).

(3) Actuators—utilize combinations of the different electro-mechanical devices such as synchronous motor, stepper motor, AC servo motor, brushless DC servo motor or brushed DC servo motor to perform some tasks.

(4) Controllers—provide necessary intelligence to control the manipulator/mobile robot, e.g., process the sensory information and compute the control commands for the actuators to carry out specified tasks.

(5) User interface—this is the part for the user to operate a robot, i.e., the control panel.

(6) Manipulator base—a robot may work in a fixed base or in a mobile base, that is to say, it can move about by employing the wheels or legs.

Fig. 10-1-1 shows an industrial robot.

Fig. 10-1-1 Industrial robot

4. Use in Industry

Today, 90% of the robots used are found in the factories and they are referred to as the industrial robots. Ten years ago, 9 out of the 10 robots were being bought by auto companies.[1] Now, only 50% of robots made today are bought by the car manufacturers. Robots are slowly finding their way into warehouses[2], laboratories, research and exploration sites, energy plants, hospitals even outer space.

The robot industry is booming, to say the least. North American robotics suppliers saw orders leap 36% in the first half of 2007[3], according to the new statistics released by Robotic Industries Association (RIA), the industry's trade group.

Robots are useful in the industry for a variety of reasons. Installing robots is often a way business owners can be more competitive[4], because robots can do some things more efficiently than people.

Robots never get sick or need to rest, so they can work 24 h a day, 7 d a week.

When the task required would be dangerous for a person, they can do the work instead.

Robots don't get bored, so work that is repetitive and unrewarding is no problem for a robot.

Although robots cannot do every type of jobs, they do well for certain industrial tasks including:

(1) Assembling operations—Assembly accounts for approximately 33% of the applications of the world robot stock (1997). Many of these robots can be found in the automotive and electronics industries.

(2) Continuous arc welding & spot welding—One of the most common uses for industrial robots is welding. Robot-welded car bodies, for example, enhance safety since a robot never misses a welding spot and performs equally well all through the day. Nearly 25% of all the industrial robots are used in different welding applications.

(3) Packaging/Palletizing-Packaging/Palletizing—It is still a minor application area for the industrial robots, accounting for only 2.8% (1997).

The following jobs are most likely to be taken over by robots in large-scale production in industry: spray coating/painting, material removal, machine loading, material transfer, cutting operations, parts inspection, parts sorting, parts cleaning, parts polishing. Fig. 10-1-2 shows applications of industrical robots.

Fig. 10-1-2 Applications of industrial robots

As the robots become cheaper, we will also see more robots taking over the jobs normally done by human beings in all walks of life.

Section Ⅱ New Words and Phrases

ascribe	*v.* 归于，归因于
serf	*n.* 农奴
coin	*vt.* 模制，借用，杜撰（新字）等
Czech	*adj.* 捷克语的 *n.* 捷克人
infrastructure	*n.* 基础设施，内部构造
sensor	*n.* 传感器
toxic	*adj.* 有毒的
actuator	*n.* 执行机构；致动器
synchronous	*adj.* 同步的
stepper	*n.* 步进（电机）
servo	*n.* 伺服
utilize	*vt.* 用，采用，利用
brushless DC servo motor	*n.* BLDC 直流无刷伺服电机
intelligence	*n.* 智能
carry out	执行
user interface	用户接口，用户界面
manipulator base	操纵台
mobile	*adj.* 移动的
auto	*n.* 汽车（automobile 的缩写）
warehouse	*n.* 仓库
exploration site	勘探现场
boom	*v.* 繁荣，火爆
to say the least	至少可以说
leap	*v.* 跳跃
supplier	*n.* 供应商
statistics	*n.* 统计数字；统计学
release	*vt.* 释放，出台 *n.* 版本，释放
repetitive	*adj.* 重复的
unrewarding	*adj.* 无偿的，不计报酬的
account for	占（多少比例）
stock	*n.* 库存，存货
continuous arc welding	连续弧焊
spot welding	点焊
miss	*v.* 错过

palletize	*v.* 码垛，堆码
minor	*adj.* 次要的
spray coating/painting	喷漆
removal	*n.* 清除
transfer	*v.* &*n.* 转移，转换
inspection	*n.* 检测，检查
polish	*v.* 抛光

Section Ⅲ Notes to Complex Sentences

[1] …9 out of the 10 robots were being bought by auto companies…

……9/10 的机器人为汽车制造公司所购买……

9 out of 10：十之有九，为习惯用语；were being bought：被动语态的过去进行式。auto：automobile 的简写。

[2] Robots are slowly finding their way into the warehouses…

机器人正慢慢进入仓库……

find way into：进入，进军。

[3] North American robot suppliers saw the orders leap 36% in the first half of 2007…

2007 上半年北美机器人供应商的订单飙升 36% ……

see…在本文中多次出现，本意是"看见"，翻译时可以灵活处理。

[4] Installing the robots is often a way business owners can be more competitive…

装配机器人往往能使公司更具竞争力……

Section Ⅳ Exercises

Translate the following paragraphs into Chinese.

End Effector

The end effector found in most robot applications is a device connected to the wrist flange of the manipulator arm. The end effector is used in many different situations in the production area, for example, it can be used for picking up parts, for welding, or for painting. The end effector gives the robotic system the flexibility necessary for operation of the robot.

The end effector is generally designed to meet the need of the robot's user. These devices can be manufactured by the robot manufacturer or by the owner of the robotic system.

The end effector is the only component found on the robotic system that may be changed from one job to another. For example, a robot can be connected to a water jet cutter, which is used to cut aside panels for auto production lines. It can be demanded to stack parts onto a tray also. In this simple process, the robot's end effector was changed, allowing the robot to be used in other application. The changing of the end effector and the reprogramming on the robot allow this system to be very flexible.

Section V Supplementary Reading

Robot

(1) Introduction.

The industrial robot and its operation is the topic of this text. The industrial robot is a tool that is used in the manufacturing environment to increase the productivity. It can be used to do routine and tedious assembly lines jobs or it can perform jobs that might be hazardous to the human workers. For example, one of the first industrial robots was used to replace the nuclear fuel rods in the nuclear power plants. A human doing this job might be exposed to the harmful amounts of radiation. The industrial robot can also operate on the assembly lines, putting together small components, such as placing the electronic components on a printed circuit board. Thus, the human worker can be relieved of the routine operation of this tedious task. Robots can also be programmed to defuse the bombs, to serve the handicapped and to perform the functions in numerous applications in our society.

The robot can be thought of as a machine that will move an end-of-arm tool, sensor and/or gripper to a preprogrammed location. When the robot arrives at this location, it will perform some sort of the task. This task could be welding, sealing, machine loading, machine unloading or a host of the assembly jobs. Generally, this work can be accomplished without the involvement of a human being, except for the programming and turning the system on and off.

(2) Robot Terminology.

A robot is a reprogrammable, multifunctional manipulator designed to move parts, materials, tools or special devices through the variable programmed motions for the performance of a variety of different tasks. This basic definition leads to other definitions, presented in the following paragraphs, which give a complete picture of a robotic system.

Preprogrammed locations are paths that the robot must follow to accomplish the work. At some of these locations, the robot will stop and perform some operation, such as assembly of parts, spray painting or welding. These preprogrammed locations are stored in the robot's memory and are recalled later for continuous operation. Furthermore, these preprogrammed locations, as well as other program data, can be changed later as the work requirements change. Thus, with regard to this programming feature, an industrial robot is very much like a computer, where data can be stored and later recalled and edited.

The manipulator is the arm of the robot. It allows the robot to bend, reach and twist. This movement is provided by the manipulator's axes, as shown in Fig. 10-5-1.

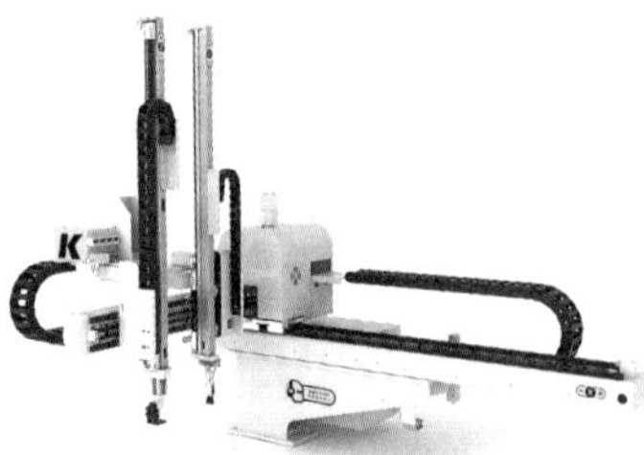

Fig. 10-5-1 Manipulator's axes

Exercises

1. Write T(True) or F(False) beside the following statements about the text.

(1) ____________ A robot is a reprogrammable, multifunctional machine designed to manipulate materials, parts, tools or specialized devices through the variable programmed motions for the performance of a variety of tasks.

(2) ___________ A robot is a manual device that performs functions naturally ascribed to humans or a machine which acts as like a human.

(3) __________ Today, 90% of all the robots used are found at home and they are referred to as the intelligent robots.

(4) __________ Installing robots is often a way business owners can be more competitive, because robots can do some things more efficiently than people.

(5) __________ The industrial robot is a tool that is used in any kind of working environment to help the convenience.

(6) ___________ Generally, this work can be accomplished without the involvement of a human being, except for the programming and turning the system on and off.

(7) __________ At every locations, the robot will not stop to perform some operation, such as assembly of parts, spray painting or welding.

(8) __________ The manipulator is the arm of the robot. It allows the robot to bend, reach and twist. This movement is provided by the manipulator's axes.

2. Choose the best answer.

(1) Installing robots is often a way business owners can be more __________, because robots can do some things more efficiently than people.

A. useful B. practical C. popular D. competitive

(2) The industrial robot is a tool that is used in the manufacturing environment to increase

the ________.

A. convenience B. efficiency C. production D. productivity

(3) The industrial robot can also operate on the ________, putting together small components, such as placing the electronic components on a printed circuit board.

A. factory B. assembly line C. laboratory D. imagination

(4) Robots can also be programmed to defuse the bombs, to serve the handicapped and to perform the functions in numerous________ in our society.

A. purposes B. practices C. adoptions D. applications

(5) Furthermore, these ________locations, as well as other program data, can be changed later as the work requirements change.

A. preparatory B. pre-made C. preserved D. preprogrammed

3. Fill in the blanks with words or expressions according to the text.

(1) Robot-welded car bodies for example enhance________ since a robot never misses a ________ and performs equally well all through the day.

(2) The end effector gives the robotic system the________ necessary for ________of the robot.

(3) It can be used to do routine and tedious________ jobs or it can perform jobs that might be________ to the human workers.

(4) This basic ________ leads to other definitions, presented in the following paragraphs, which give a complete picture of a________ system.

(5) Furthermore, these________ as well as other________ can be changed later as the work requirements change.

(6) The________ is the arm of the robot. It allows the robot to bend, reach and twist. This movement is provided by the manipulator's________.

4. Answer the following questions according to the text.

(1) 什么是机器人？机器人的组成有什么？机器人有什么工业应用？

(2) 什么是工业机器人？工业机器人的工作原理是什么？

(3) 机器人是否是计算机？它是如何对位置进行预编程的？

5. Translate the following paragraphs into Chinese.

(1) A robot is a reprogrammable, multifunctional manipulator designed to move parts, materials, tools or special devices through the variable programmed motions for the performance of a variety of different tasks. This basic definition leads to other definitions, presented in the following paragraphs, which give a complete picture of a robotic system.

(2) Robots are useful in the industry for a variety of reasons. Installing robots is often a way business owners can be more competitive[4], because robots can do some things more efficiently than people.

(3) Furthermore, these preprogrammed locations, as well as other program data, can be changed later as the work requirements change. Thus, with regard to this programming feature, an industrial robot is very much like a computer, where data can be stored and later recalled and edited.

单元评价

通过本单元课文阅读，单词、短语练习，扩展阅读及习题练习，使学生能够掌握与工业机器人及其应用领域相关的词汇和表达语句，能够自行完成相关英文的翻译，并具备与技术人员利用英语进行工业机器人相关内容交流的能力。同时通过单元学习，培养学生独立学习、灵活运用所学知识分析并解决问题的能力。

单元 11

机械产品质量检测与装配

【学习目标】

知识目标：

1. 理解与掌握关于测量的基本原理与方法的相关英语词汇。

2. 掌握互换性、公差及标准的基本英语知识及常用的英语表达语句。

技能目标：

1. 具备机械产品测量相关英语词汇跟读技巧，能够运用英语识读三坐标测量仪操作说明。

2. 具备使用英语描述机械产品测量相关知识的能力。

素养目标：

1. 培养学生自我提升、开拓创新的能力。

2. 培养学生创新的思维能力以及严谨求实的学习态度。

Section Ⅰ Texts

1. Product Quality Inspection

Product quality inspection refers to the technical activities of inspecting which tests and measures one or more quality characteristics of a product according to the product standards or inspection procedures and compares the results with the specified quality requirements to determine the conformity of each quality characteristic.

According to the different requests for the utilization of the products, each product has its own quality characteristics.[1] These characteristics are generally translated into the specific quality requirements in the product technical standards (national standards, industry standards, enterprise standards) and other relevant product design drawings, process and manufacturing technical documents, as the basis for quality inspection and the comparison for the post-inspection test results of the reference base. In order to ensure the quality of the product, it is necessary to conduct the quality inspection on raw materials, purchased parts, outsourced parts, blank, semi-finished products, finished products, etc, in the production process and strict control, so that the unqualified raw materials are not put into the production, unqualified semi-finished products are not transferred, unqualified parts are not assembled and the unqualified products are not delivered. The reputation of the producers has to be maintained and the social benefits have to be improved. Product quality inspection (Fig. 11-1-1) is an important part of the quality management in the production. [2]

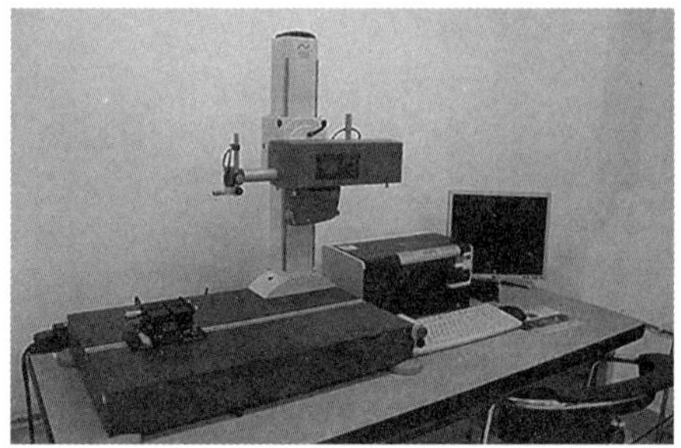

Fig. 11-1-1 Product quality inspection

2. Machine Assembly

Mechanical products are generally composed of many parts and components. According to the specified technical requirements, the process of combining several parts into the module, components, or components and several parts into the products is called assembly.

Mechanical assembly(Fig. 11-1-2) is the last stage in the whole process of the mechanical manufacturing, which occupies a very important position in the manufacturing process.[3] The quality of the mechanical products is ultimately guaranteed by the assembly work. Parts quality is the basis of mechanical product quality, but the assembly process is not a simple combination of the qualified parts. Even if the high-quality parts are available, a low-quality assembly still may produce a low-quality product. With the economical precision parts and components, high quality products can be assembled through the high quality assembly.[4]

Fig. 11-1-2 Mechanical assembly

In recent years, due to the rapid improvement of the mechanization and automation in blank manufacturing and mechanical processing, there is a tendency that the proportion of the assembly workload in the manufacturing process to expand. Therefore, it is necessary to improve the technical level and labor productivity of the assembly work, in order to adapt to the development trend of the whole engineering industry.

For products with relatively complex structure, in order to ensure the assembly quality and efficiency, the product should be decomposed into the assembly units for separate assembly according to the structural characteristics of the product.[5]

Parts are the most basic units of the mechanical products, the parts are generally assembled into the combinations, components or modules after the assembly is made into the machine. A combination is made of several parts permanently connected or connected and then processed. A component is a combination of the parts and combinations. Parts in the machine can complete a certain and complete function.

3. Coordinate Measuring Machine

Three axes of coordinate measuring machine (CMM) (Fig. 11-1-3) are equipped with the air source brake switch and micro device, which can realize the single axis precision transmission and adopt the high performance data acquisition system. CMM is applied into the product design, mold equipment, gear measurement, blade measurement, machine manufacturing, work fixture, electronic appliance and other precision measurement.

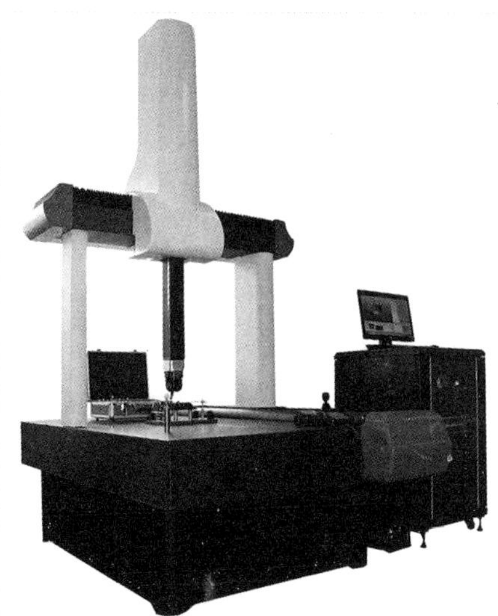

Fig. 11-1-3 Coordinate measuring machine

CMM is a kind of instrument that can show the measuring ability of the geometric shape, length and circumference in the hexahedron space range, also known as the CMM or CMM measuring bed. Three CMM can also be defined as "a detector that can move in three directions and it can move on three mutually perpendicular guide rails. The detector transmits the signals by the contact or non-contact and the displacement measuring system of three axes (such as grating ruler) is calculated by a data processor or computer to measure the points (x, y, z) and various functions of the workpiece". The measuring functions of CMM should include the dimensional accuracy, positioning accuracy, geometric accuracy and contour accuracy. It is widely used in the measurement of the box, frame, gear, cam, worm gear, worm, blade, curve, surface and in the fields of the automobile, electronics, machinery, aviation, military, mold and other industries.

In simple terms, CMM is a measuring equipment that possesses the guiding mechanism, length measuring element and digital display device in the three direction which is mutually perpendicular. Besides, there is a table to place the workpiece. The probe can move lightly to the measured point manually or mechanically and display the measured point coordinate value by reading the equipment and digital display device. This is clearly the simplest and most primitive measuring machine. With this measuring machine, the coordinate value at any point in the measured volume can be displayed by reading the digital display device. The point picking and sending device of the measuring machine is the probe, which is equipped with grating ruler and reading head along the direction of x, y and z axes. The measuring process is that the control system collects the coordinate value of the current three-axis coordinate relative to the origin of the machine tool, and then the computer system processes the data, when the probe contacts the workpiece and sends out the mining point signals.[6]

Section Ⅱ New Words and Phrases

inspection	*n*. 视察；检查，审视
product quality inspection	产品质量检验
technical activities	技术活动
characteristics	*n*. 特性，特征；特色；特质
according to	*prep*. 根据，依照（原则等）
quality	*n*. 质量，品质；*adj*. 优质的，高质量的
determine	*v*. 决定，控制；查明，确定
conformity	*n*. 遵守，依照；符合，一致
technical standards	技术标准
national standards	国家标准
industry standards	行业标准
enterprise standards	企业标准
drawings	*n*. 图纸；图示；提用
technical documents	技术文件
ensure	*v*. 确保，保证；保护，使安全
necessary	*adj*. 必要的，必需的；必然的
conduct	*v*. 实施，进行；*n*. 行为，举止

purchased part	外购件
outsourced part	外协件
blank	*n.*（金属或木头的）毛坯件，毛坯料；*adj.* 空白的，空的
production	*n.* 生产，制造；（对自然物的）加工，采集
transfer	*v.*（使）转移，搬迁；*n.* 转移，转让，调动
assemble	*v.*（使）集合，（使）聚集；装配，组装
deliver	*v.* 投递，运送；履行，兑现
interests	*n.* 利益（interest 的复数）
consumer	*n.* 消费者，消耗者；用户，顾客
reputation	*n.* 名誉，名声
benefits	*n.* 利益，好处；福利
machine assembly	机器装配
compose	*v.* 组成，构成
several	*adj.* 不同的，各自的；*pron.* 几个，数个
occupy	*v.* 使用，居住；占据（空间，时间）
available	*adj.* 可用的，可获得的；
rapid	*adj.* 快的，迅速的；（移动，活动）快速的
certain	*adj.* 必然的，必定的，确定的 *pron.* 某些，某几个
coordinate measuring machine（CMM）	三坐标测量仪
dimensional accuracy	尺寸精度
positioning accuracy	定位精度；定位准确度
geometric accuracy	几何精度
contour accuracy	轮廓精度
box	*n.* 盒，箱，匣；箱体
frame	*n.* 框架，边框；构架，支架，机架
gear	*n.* 排挡，齿轮
cam	*n.* 凸轮
worm gear	蜗轮；螺旋齿；蜗轮蜗杆
blade	*n.* 刀片，刀刃；（船或飞机螺旋桨的）桨叶
curve	*n.* 曲线，弧线；转弯，弯道
surface	*n.* 曲面，表面，水面，地面；桌面，台面
automobile	*n.* 汽车
electronics	*n.* 电子学；电子设备，电子器件
machinery	*n.* 机器，机械（尤指大型机械）；装置
aviation	*n.* 航空，飞行（术）；航空工业，飞机工业

Section Ⅲ Notes to Complex Sentences

[1] According to the different requests for the utilization of the products, each product has its own quality characteristics.

根据产品的使用要求不同，每种产品都有各自的质量特性。

[2] In order to ensure the quality of the product...The reputation of the producers has to be maintained and the social benefits have to be improved. Product quality inspection is an important part of quality management in production.

为确保产品质量……应维护生产者信誉和提高社会效益。产品质量检测是生产中质量管理的一个重要组成部分。

[3] Mechanical assembly is the last stage in the whole process of the mechanical manufacturing, which occupies a very important position in the manufacturing process.

机械装配是整个机械制造过程中的最后一个阶段，在制造过程中占有非常重要的地位。

[4] Even if the high-quality parts are available, a low-quality assembly still may produce a low-quality product. With the economical precision parts and components, high quality products can be assembled through the high quality assembly.

即使使用高质量的零部件，低质量的装配也可能装出低质量的产品；高质量的装配则可以在经济精度零件的基础上，装配出高质量的产品。

[5] For products with relatively complex structure, in order to ensure assembly quality and efficiency, the product should be decomposed into assembly units for the separate assembly according to the structural characteristics of the product.

对于结构比较复杂的产品，为了保证装配质量和装配效率，需要根据产品的结构特点，从装配工艺角度将产品分解为单独进行装配的装配单元。

[6] The measuring process is that the control system collects the coordinate value of the current three-axis coordinate relative to the origin of the machine tool, and then the computer system processes the data, when the probe contacts the workpiece and sends out the mining point signals.

其测量过程就是当测头接触工件并发出采点信号时，由控制系统去采集当前机床三轴坐标相对于机床原点的坐标值，再由计算机系统对数据进行处理。

Section Ⅳ Exercises

Translate the following paragraph into Chinese.

Mechanical product quality inspection is mainly about the quality characteristics and standards of the mechanical products, then compare the testing results to the prescribed standards to judge whether the mechanical products meet the standard. The mechanical product quality inspection can not only realize the beneficial monitoring of the product quality, but can also contribute to the product quality management and other aspects. Products that had been tested for the quality has the conditions for market works, which makes the product quality inspection play an indispensable role in the mechanical product inspection process. In order to increase the competitiveness of the mechanical products manufacturing enterprises in the market, among the mechanical products quality inspection, the mechanical product inspectors should be clear about the starting position, prepare the required tools adequately and select the appropriate inspection means based on the actual situation to avoid the appearance of the unqualified products which affect the enterprise reputation and market field competitiveness.

Section V Supplementary Reading

Mechanical Product Testing Procedures

1. Master the Characteristics and Mechanical Properties of the Mechanical Products

Different uses make the mechanical products themselves have different structures and different small parts distributed in various parts of the mechanical products. Each part has different functions and uses. Nowadays, each mechanical product needs to adopt different methods to transfer load, so every part of the mechanical products needs to have some effects, which requires each part to be able to bear external forces and give play to the advantages of mechanical properties and organizational structure. In addition, every part of a mechanical product needs to be resistant to the wearing out, corrosion and being interchangeable. Therefore, we should fully understand the characteristics and mechanical properties of the mechanical products through the inspection of the mechanical products.

2. Analyze the Quality Characteristics of the Mechanical Products

Mechanical products should be inspected by quality items and quality methods, and the quality characteristics of the mechanical products are the basis of these inspection conditions. It is very important to fully understand the quality characteristics of the mechanical products in the process of the quality inspection. Because the parts of the mechanical products have different functions, the parts of each mechanical products have different characteristics. For example, the parts of the mechanical products have the chemical and mechanical properties, different structure sizes, geometric parameters and the surface roughness are not the same. The parts of some mechanical products also have many quality characteristics, such as the functional index, structural size, connection degree of moving the parts and service life. Due to the large number of parts and different functions, the quality characteristics of the mechanical products should be fully analyzed.

3. Select the Method for a Mechanical Product test

The inspection of the mechanical products can be carried out by the total inspection method and sampling inspection method. Different product quantities choose different methods. Compared with the sampling method, the total inspection is more suitable for the quality inspection of the mechanical products with large quality problems, of which the measurement is convenient and the cost is lower. The sampling inspection rule is to inspect the mechanical products according to the sampling table of the sampling inspection, which is mostly used in the mass-produced mechanical products and some items with the destructive inspection requirements.

Exercises

1. Write T(True) or F(False) beside the following statements about the text.

(1) __________ According to the different requests for the utilization of the products, each product has its own quality characteristics.

(2) __________ The reputation of the producers is not necessary to be maintained and the social benefits may not be improved.

(3) __________ According to the specified technical requirements, the process of combining several parts into the module, components, or components and several parts into the products is called assembly.

(4) __________ Mechanical assembly is just at the starting stage in the whole process of the production, which occupies a very important position in the production process.

(5) __________ It is only used in the measurement of the box, frame, gear, cam, worm gear, worm, blade, curve, surface and in other industries.

(6) __________ The measuring process is that the control system collects the coordinate value of the current three-axis coordinate relative to the origin of the machine tool, and then the computer system processes the data, when the probe contacts the workpiece and sends out the mining point signal.

(7) __________ All the uses make the mechanical products themselves have different structures and same small parts distributed in all the parts of the mechanical finished products.

(8) __________ Therefore, we should fully understand the characteristics and mechanical properties of the mechanical products through the inspection of the mechanical products.

2. Choose the best answer.

(1) According to the different requests for the __________ of the products, each product has its own quality characteristics.

A. adoption B. employment C. utilization D. embracing

(2) Product quality __________ is an important part of the quality management in the production.

A. check-in B. inspection C. detection D. probing

(3) Therefore, it is necessary to improve the technical level and labor productivity of the

assembly work, in order to adapt to the development trend of the whole ________ industry.

A. medical　　B. civil works　　C. construction　　D. engineering

(4) Parts quality is the basis of mechanical product quality, but the ________ process is not a simple combination of the qualified parts. Even if the high-quality parts are available, a low-quality assembly still may produce a low-quality product.

A. assembly　　B. production　　C. manufacturing　　D. construction

(5) CMM is a kind of ________ that can show the measuring ability of the geometric shape, length and circumference in the hexahedron space range, also known as the CMM or CMM measuring bed.

A. tool　　B. equipment　　C. instrument　　D. appliance

3. Fill in the blanks with words or expressions according to the text.

(1) In simple terms, CMM is a ________ that possess the guiding mechanism, length measuring element and digital display device in the three direction which is mutually ________.

(2) The measuring process is that the control system collects the coordinate value of the current ________ relative to the origin of the machine tool, and then the computer system processes the data, when the probe contacts the workpiece and sends out the ________.

(3) Different uses make the ________ themselves have ________ and different small parts distributed in various parts of the mechanical products.

(4) Therefore, we should fully understand the ________ and mechanical ________ of the mechanical products through the inspection of the mechanical products.

(5) It is very important to fully understand the ________ of the mechanical products in the process of the ________.

(6) The sampling inspection rule is to inspect the mechanical products according to the sampling table of the ________, which is mostly used in the mass-produced mechanical products and some items with the ________ requirements.

4. Answer the following questions according to the text.

(1) 什么是产品质量检测？为什么要进行产品质量检测？

(2) 什么是机械装配？机械装配有什么重要的作用？

(3) 三坐标测量仪的工作原理是什么？机械产品检测的步骤是什么？

5. Translate the following paragraphs into Chinese.

(1) Product quality inspection refers to the technical activities of inspecting which tests and measures one or more quality characteristics of a product according to the product standards or inspection procedures and compares the results with the specified quality requirements to determine the conformity of each quality characteristic.

(2) Mechanical products are generally composed of many parts and components. According to the specified technical requirements, the process of combining several parts into the module, components, or components and several parts into the products is called assembly.

(3) Three axes of CMM are equipped with the air source brake switch and micro device, which can realize the single axis precision transmission and adopt the high performance data acquisition system. CMM is applied into the product design, mold equipment, gear measurement, blade measurement, machine manufacturing, work fixture, electronic appliance and other precision measurement.

单元评价

通过本单元课文阅读，单词、短语练习，扩展阅读及习题练习，使学生掌握关于测量基本原理与方法相关的词汇和表达语句，能够自行识读三坐标测量仪说明书，并具备与技术人员利用英语进行机械产品测量及检测相关内容交流的能力，同时培养学生的创新思维能力以及严谨求实的学习态度。

单元 12

逆向工程与 3D 打印

【学习目标】

知识目标：

1. 熟悉逆向工程与 3D 打印相关的英语词汇。
2. 掌握与技术人员交流逆向扫描、创新设计等时常用的英语表达语句。

技能目标：

1. 具备逆向扫描和快速成形技术相关英语词汇跟读技巧，能够运用英语检索 3D 打印相关文献及技术资料。
2. 具备使用英语描述逆向设计和 3D 打印的能力。

素养目标：

1. 加强学生安全生产、环境保护、节约资源的意识。
2. 增强学生的钻研精神和创新意识。
3. 培养学生刻苦钻研能力，开发学生的创新思维能力。

Section I Texts

1. Reverse Engineering

Reverse engineering (RE), also known as the reverse technology, is a product design technology, reproduction process, reverse analysis and research of a target product, thus, the product's processing, organizational structure, functional characteristics and technical specifications and other design elements are concluded. Sequentially, the products possess the similar function but not identical. Reverse engineering comes from hardware analysis in the commercial and military fields. Its main purpose is to deduce the design principle of the product directly from the analysis of the finished product without easy access to the necessary production information.[1] The implementation process of the reverse engineering is a collaborative process of many fields and disciplines.

In the general concept of the engineers and technicians, the product design process is a process from design to product.[2] The designers first conceive the shape, performance and general technical parameters of the product in the brain, then, complete various data models in the detailed design stage and, finally, transfer this model to the research and development process to complete the whole design and development cycle of the product. This product design process is called "forward design" or "top down design" process. Reverse engineering product design can be considered as a process from the product to design. Fig. 12-1-1 shows the procedure of reverse degign. Simply, reverse engineering product design is the process of the reverse product design data (including various design drawings or data models) based on the existing products. In this sense, the reverse engineering has been used in industrial design for a long time. For example, the hull lofting design commonly used in the early shipbuilding industry is a good example of the reverse engineering. Fig. 12-1-2 shows a product of reserse design.

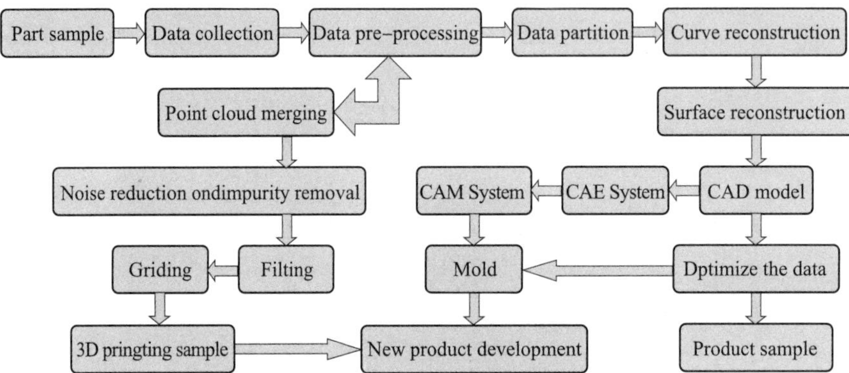

Fig. 12-1-1 A procedureof reverse design

Fig. 12-1-2 A product of reverse design

2. Effect of the Reverse Engineering

Reverse engineering is widely used in the new product development, product modification design, product imitation, quality analysis, testing and other fields.[3] Its functions are as follows:

(1) Shorten the product design, development cycle, accelerate the product replacement speed;

(2) Reduce the cost and risk of developing the new products;

(3) Accelerate the product modeling and serialization design;

(4) Suitable for the single and small batch parts manufacturing, especially mold manufacturing, which can be divided into direct and indirect mold manufacturing method.

3. Reverse Scanning

The so-called reverse scanning (Fig. 12-1-3) technology is to conduct 3D scanning and data collection on the original shape of the real object, constructing 3D models with the same shape structure after data processing and 3D reconstruction. The purpose of the reverse scanning is to obtain point cloud by the physical objects and optimize the design and innovative design based on the point cloud.[4] The three-dimensional display of the reverse scanning technology can scan the structure of the objects in multiple directions by the software, then establish the three-dimensional digital models of objects.

The principle reverse scanning is a combination of the structured light technology, phase measurement technology, 3D vision technology and composite three-dimensional non-contact measurement technology. So it is also called "three-dimensional structured light scanner". The use of the 3D scanning technology makes it possible to take photographic measurements of the objects. The so-called photographic measurement is similar to taking photographs by the camera of the objects in the field of vision. The difference is that the camera takes two-dimensional images of the objects, while the developed measuring instrument acquires three-dimensional information of the objects. Unlike the traditional three-dimensional scanners, the scanner can measure a face at the same time.[5]

Fig. 12-1-3 Reverse scanning

4. 3D Printing

3D printing (3DP) (Fig. 12-1-4) is a kind of rapid prototyping technology, also known as additive manufacturing.[6] It is a kind of technology that uses the powder metal or plastic and other adhesive materials to construct objects based on digital model files through the way of printing layer by layer.

3D printing is usually realized by using the digital material printers. It is often used to make the models in mold manufacturing, industrial design and other fields. Then, it is gradually used in the direct manufacturing of some products. There are already the parts printed by using this technology. The technology is used in the jewelry, footwear, industrial design, architecture, engineering and construction, automotive, aerospace, dental and medical industries, education, geographic information systems, civil engineering, firearms and other fields.

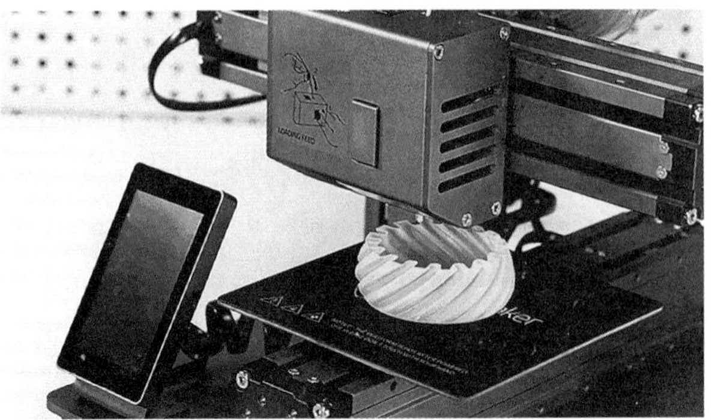

Fig. 12-1-4 3D printing

Ordinary printers used in daily life can print flat objects designed by the computers, while the so-called 3D printers work basically the same principle as the ordinary printers, except for some different printing materials. The printing materials of the ordinary printers are ink and paper, while the 3D printers are equipped with the metal, ceramic, plastic, sand and other "printing materials", which are real raw materials. When connected to a computer, the printer can be controlled by the computer to add layer upon layer of the "print material", eventually turning the blueprints on the computer into the real objects. A 3D printer is a device that can "print" real 3D objects, such as a robot, a toy car, models and even food. It is colloquially called "printers" in reference to the technical principles of the ordinary printers, because the process of the layering is very similar to that of inkjet printing. The printing technology is called the 3D stereoscopic printing.

There are many different techniques for the 3D printing. They differ in creating the parts in the manner of the available materials and in different layers of the construction. Common 3D printing materials include the nylon glass fiber, durable nylon material, gypsum material, aluminum material, titanium alloy, stainless steel, silver plated, gold plated and rubber materials.

Section Ⅱ New Words and Phrases

reverse engineering	逆向工程
organizational structure	组织结构
functional characteristic	功能特征
technical specification	技术规格
element	*n.* 要素；元素
function	*n.* 功能，职责；*v.* 工作，运转
identical	*adj.* 完全相同的；同一的

hardware	*n.* 硬件；五金制品；装备，设备
commercial	*adj.* 商业的，商务的；商业化的
military	*adj.* 军事的，军队的
implementation	*n.* 实现，实施，执行
concept	*n.* 概念，观念
performance	*n.* 性能 *adj.* 性能卓越的，高性能的
parameter	*n.* 决定因素，规范；参数
detailed design stage	详细设计阶段
forward design	正向设计
hull lofting	船体放样设计
new product development	新产品开发
product modification design	产品改型设计
product imitation	产品仿制
quality analysis and testing	质量分析与检测
reverse scan	逆向扫描
obtain	*v.* 得到，获得
point cloud	点云
three-dimensional	*adj.* 三维的；立体的；真实的
three-dimensional structured light scanner	三维结构光扫描仪
3D printing（3DP）	3D 打印
rapid prototyping technology	快速成形技术
additive manufacturing	增材制造
digital material printer	数字技术材料打印机
ordinary	*adj.* 普通的，平常的；*n.* 常见的人（或事物）(the ordinary)
metal	*n.* 金属，合金　*v.* 用金属做　*adj.* 金属制的
ceramic	*n.* 陶瓷制品，陶瓷器；制陶艺术 *adj.* 陶瓷的
plastic	*n.* 塑料；*adj.* 塑料制的
sand	*n.* 沙，沙子；*v.*（用砂纸或打磨机）磨光
raw material	原材料
nylon glass fiber	尼龙玻璃纤维
durable nylon material	耐用尼龙材料
gypsum material	石膏材料
aluminum material	铝材料
titanium alloy	钛合金
stainless steel	不锈钢
gold /silver plated	镀金的/镀银的
rubber material	橡胶材料

Section Ⅲ Notes to Complex Sentences

[1] Its main purpose is to deduce the design principle of the product directly from the analysis of the finished product without the easy access to the necessary production information.

其主要目的是在不能轻易获得必要的生产信息的情况下，直接从成品分析，推导出产品的设计原理。

[2] In the general concept of the engineers and technicians, the product design process is a process from the design to product...

在工程技术人员的一般概念中，产品设计过程是一个从设计到产品的过程……

[3] Reverse engineering is widely used in new product development, product modification design, product imitation, quality analysis and testing and other fields.

逆向工程被广泛地应用到新产品开发和产品改型设计、产品仿制、质量分析检测等领域。

[4] The purpose of the reverse scanning is to obtain point cloud by the physical objects and optimize the design and innovative design based on point cloud.

逆向扫描的目的是利用实物获取点云，并基于点云进行优化设计以及创新设计。

[5] Unlike traditional three-dimensional scanners, the scanner can measure a face at the same time.

与传统的三维扫描仪不同的是，该扫描仪能同时测量一个面。

[6] 3D printing (3DP) is a kind of rapid prototyping technology, also known as the additive manufacturing...

3D 打印（3DP）即快速成形技术的一种，又称增材制造……

Section Ⅳ Exercises

Translate the following paragraphs into Chinese.

Reverse engineering technology is a comprehensive application of the modern industrial design theory, method, production engineering, material engineering, related professional knowledge, systematic analysis and research. And, then, rapidly develop and manufacture new products with high added value and high technical level. The combination of the technology and rapid prototyping technology can realize the rapid three-dimensional copy of the products and after the CAD modeling modification or rapid prototyping process parameters adjustment, it can also realize the variation recovery of the parts or models.

For a long time, the traditional way of the industrial product development is from the conception of the product requirements, the determination of the function and specification expectations, to the design, manufacturing, assembling and performance testing of each component.

Each component retains the original design, this development pattern is called "Forward Engineering".

However, with the improvement of the industrial technology level and the living standards, any universal products in the consumer's requirements for the high quality and the functional demand are no longer the only condition to win the market competitiveness. Therefore, another important route in the process of new product development is to reverse the sample. Reverse engineering technology is also called reverse engineering.

Section V Supplementary Reading

Rapid Prototyping

Rapid prototyping (RP) technology was born in the late 1980 s and is regarded as a major achievement in the manufacturing field in the past 20 years. It integrates the mechanical engineering, CAD, reverse engineering technology, layered manufacturing technology, numerical control technology, material science and laser technology, which can automatically, directly, quickly and accurately transform the design idea into a functional prototype or direct manufacturing parts, so as to provide a high efficiency and low cost implementation means for the parts prototyping, new design idea verification and other aspects.

1. Overview of the Rapid Prototyping Technology

After the 1990s, the external situation of the manufacturing industry changed fundamentally. The personalized and variable demand of the users forced the enterprises to gradually abandon the original "scale economy first" as the characteristics of the less varieties and mass production mode. Then, the modern production mode of multi-varieties and small batch were adopted, according to the order organization.

At the same time, the globalization and integration of the market require more enterprises to have a high degree of the sensitivity. In the face of the rapidly changing market environment, new products are constantly developed which change the passive to adapt to the user for the active guide of the markct and guarantee that the enterprise in an invincible position is in the competition.

In this background, it is visible that the focus of the market competition transfers to speed and higher performance/price products of the enterprise, which will help to have a stronger comprehensive competitiveness.

Rapid prototyping technology is an important branch of the advanced manufacturing technology, which has made great breakthroughs in both manufacturing ideas and implementation methods. Rapid prototyping technology can be used to rapidly evaluate and modify the product design, automatically and rapidly transform the design into the prototype products with the corresponding structure and function or directly manufacture parts, thus greatly shortening the development cycle of the new products and reducing the development cost of products. So, the enterprises can quickly respond to the market demand, improve the market competitiveness of products and the comprehensive competitiveness of the enterprises.

2. Application

Constantly improving the application level of RP technology is an important aspect to promote the development of RP technology. school of Mechanical Engineering, Xi'an Jiaotong University (XJTU), National Engineering Research Center of Rapid Prototyping and Ministry of Education Rapid Prototyping engineering research center have widely used RP in the industrial modeling, machine manufacture, aerospace, military, architecture, household appliances, light industry, medicine, archaeology, culture, arts, sculpture, jewelry and other fields. And with the development of the technology itself, its applications will continue to expand. The practical application of RP technology mainly focuses on the following aspects.

Application in the process of new product modeling design: Rapid prototyping technology has established a new product development mode for industrial product designers. The application of RP technology can quickly, directly and accurately transform the design idea into a functional object model, which not only shortens the development cycle, but also reduces the development cost and enables the enterprise to take the advantage in the fierce market competition.

Application in the field of machinery manufacturing: Characteristics of RP technology itself make it obtain a wide range of applications in the field of machinery manufacturing, mostly used in the manufacture of single piece, small batch metal parts manufacturing. Some special complex parts, need only single production, or less than 50 pieces of small batch, generally available RP technology directly for molding, low cost, short cycle.

Application in rapid mold manufacturing: Traditional mold production time is long and the cost is high. Combining the rapid prototyping technology with the traditional mold manufacturing technology can greatly shorten the development cycle of the mold manufacturing and improve the productivity, which is an effective way to solve the weak link of mold design and manufacturing. The application of the rapid prototyping technology in the mold manufacturing can be divided into two kinds, direct mold making and indirect mold making. Direct mold making refers to the use of the RP technology to directly pile up the mold and indirect mold making is to produce rapid prototyping parts first, and then copy the required mold by the parts.

Application in medical field: There are many researches on the application of RPM technology in the medical field. Based on medical image data, using RP technology to make human organ model has the great application value to the surgery.

Application in the field of culture and art: In the field of culture and art, rapid prototyping manufacturing technology is mostly used in the artistic creation, cultural relic reproduction, digital sculpture, etc.

Applications in the field of aerospace technology: In the field of aerospace, aerodynamic ground simulation experiments, namely wind tunnel tests, are an essential part of the design of the advanced performance space shuttle system (i. e. space shuttle). The model used in this experiment has complex shape, high precision requirements and streamline characteristics. RP technology is adopted, and the solid model is completed automatically by RP equipment according to the CAD model, which can guarantee the model quality well.

Application in the home appliance industry: Rapid prototyping system in the domestic home appliance industry has been widely used, so that many home appliance enterprises get in the forefront of the market, have adopted rapid prototyping system to develop new products and received good results.

Rapid prototyping technology is widely used. It is believed that with the continuous maturity and improvement of the rapid prototyping manufacturing technology, it will be promoted and applied in more and more fields. Fig. 12-5-1 shows a product of rapid prototyping.

Fig. 12-5-1 Product of rapid prototyping

Exercises

1. Write T(True) or F(False) beside the following statements about the text.

(1) ____________ Reverse engineering (RE, also known as the reverse technology) is a product design technology, reproduction process, reverse analysis and research of a target product, thus, the product's processing, organizational structure, functional characteristics and technical specifications and other design elements are concluded. Sequentially, the products possess the similar function but not identical.

(2) ____________ In the general concept of the engineering designer, the product design process is a process from production to the application.

(3) ____________ Reverse engineering product design can be considered as a plan from the design to the production.

(4) ____________ Reverse engineering is widely used in the new product development, product modification design, product imitation, quality analysis, testing and other fields.

(5) ____________ The purpose of the reverse scanning is to obtain point cloud by the planned

objects and realizing the design and polishing design based on the point cloud.

(6) __________ The use of the 3D scanning technology makes it impossible to take photographic detection of the objects.

(7) __________ In this sense, the reverse engineering has been used in industrial design for a long time.

2. Choose the best answer.

(1) Reverse engineering comes from __________ in the commercial and military fields.

A. software detection B. hardware detection

C. software analysis D. hardware analysis

(2) The purpose of the reverse scanning is to obtain point cloud by the physical objects and __________ the design and innovative design based on the point cloud.

A. shape B. polish C. improve D. optimize

(3) The purpose of the reverse scanning is to obtain __________ by the physical objects and optimize the design and innovative design based on the point cloud.

A. point cloud B. point measurement

C. spot cloud D. spot measurement

(4) Then, the modern production mode of __________ and small batch were adopted, according to the order organization.

A. multi-angles B. multi-varieties C. multi-sorts D. multi-categories

(5) They differ in creating the parts in the manner of the __________ materials and in different layers of the construction.

A. available B. obtainable C. attained D. accessible

3. Fill in the blanks with words or expressions according to the text.

(1) Its main purpose is to deduce the __________ of the product directly from the analysis of the __________ without easy access to the necessary production information.

(2) Simply, __________ is the process of the reverse product design data (including various design drawings or data models) based on the __________.

(3) The three-dimensional display of the __________ can scan the structure of the objects in multiple directions by the software, then establish the __________ of objects.

(4) It is a kind of technology that uses the powder metal or plastic and other __________ materials to construct objects based on digital model files through the way of __________ layer by layer.

(5) The __________ of the users forced the enterprises to gradually abandon the original "scale economy first" as the characteristics of the __________.

(6) It is believed that with the continuous __________ of the rapid prototyping manufacturing technology, it will be __________ in more and more fields.

4. Answer the following questions according to the text.

(1) 什么是逆向工程？逆向工程的作用是什么？如何进行逆向扫描？

(2) 什么是 3D 扫描？3D 扫描有什么重要的作用？

(3) 快速成形技术是什么？快速成形技术有哪些应用？

5. Translate the following paragraphs into Chinese.

(1) In this background, it is visible that the focus of the market competition transfers to speed and higher performance/price products of the enterprise, which will help to have a stronger comprehensive competitiveness.

(2) There are many researches on the application of RPM technology in the medical field. Based on medical image data, using RP technology to make human organ model has the great application value to the surgery.

(3) Rapid prototyping system in the domestic home appliance industry has been widely used, so that many home appliance enterprises get in the forefront of the market, have adopted rapid prototyping system to develop new products and received good results.

单元评价

通过本单元课文阅读，单词、短语练习，扩展阅读及习题练习，使学生能够掌握与逆向工程和 3D 打印相关的词汇和表达语句，能够自行完成相关文献内容的翻译，并具备与技术人员进行英语基础交流的能力，同时通过本单元的学习，加强学生安全生产、环境保护、节约资源的意识，增强学生的钻研精神和创新意识。

单元 13

智能制造生产线及应用

【学习目标】

知识目标：

1. 熟悉智能生产线及其应用的相关英语词汇。

2. 掌握与技术人员交流智能生产线、柔性制造系统时常用的英语表达语句。

技能目标：

1. 具备智能生产线相关英语词汇跟读技巧，能够熟练翻译关于智能生产线及其应用的英语描述。

2. 具备使用英语描述智能生产线的能力。

素养目标：

1. 养成综合分析、全面考虑的习惯。

2. 树立正确地分析问题、解决问题的思想，激发创新思维。

3. 形成科学严谨、一丝不苟的工作作风。

Section I Texts

An intelligent manufacturing production line integrates all major elements of the manufacturing to a highly automated system. First applied in the late 1960s, the intelligent manufacturing production line consists of a number of manufacturing cells, for which each contains an industrial robot (serving several CNC machines) and an automated material-handling system. And, they are all interfaced with a central computer. Through the workstation, different computer instructions for the manufacturing process can be downloaded for each successive part.

This system is highly automated and is capable of optimizing each step of the total manufacturing operation. These steps may involve one or more processes and operations (such as machining, grinding, cutting, forming, powder metallurgy, heat treating, and finishing) as well as the handling of the raw materials, inspection and assembly.[1] The most common applications of the intelligent manufacturing production line to date have been in the machining and assembly operations. A variety of the intelligent manufacturing production line technology is available from the machine to tool manufacturing.

Intelligent manufacturing production line systems represent the highest level of the efficiency, sophistication and productivity that has been achieved in the manufacturing plants. The flexibility of the intelligent manufacturing production line is such that it can handle a variety of part configurations and produce them in any order.

1. Elements of the Intelligent Manufacturing Production Line

The basic elements of a flexible manufacturing system are:

(1) workstations;

(2) automated handling and transport of the materials and parts;

(3) control systems.

The workstations are arranged to yield the greatest efficiency in production, with an orderly flow of the materials, parts and products through the system.

The types of the machines in the workstations depend on the type of production. For the machining operations, they usually consist of a variety of 3-5 axis machining centers, CNC lathes, milling machines, drill presses and grinders. Other various equipment is also included, such as devices for the automated inspection, assembly and cleaning.

Other types of operations suitable for the intelligent manufacturing production line include sheet metal forming, punching, shearing and forging; they incorporate furnaces, forging machines, trimming presses, heat-treating facilities and cleaning equipment.

Because of the flexibility of the intelligent manufacturing production line, the material-handling, storage and retrieval systems are very important. Material handling is controlled by a central computer and performed by automated guided vehicles, conveyors and various transfer mechanisms. The system is capable of transporting the raw materials, blanks and parts in various stages of the completion to any machine (in random order) and at any time.[2] Prismatic parts are usually moved on specially designed pallets. Parts having rotational symmetry (such as those for turning operations) are usually moved by mechanical devices and robots.

The computer control system of the intelligent manufacturing production line is its brain and includes various software and hardware. This sub-system controls the machinery and equipment in the workstations and the transporting of the raw materials, blanks, and parts in various stages of completion from machine to machine. It also stores data and provides communication terminals that display the data visually.

2. Scheduling

Because the intelligent manufacturing production line (Fig. 13-1-1) involves a major capital investment, the efficient machine utilization is essential, that is to say, the machines must not stand idle. Consequently, proper scheduling and process planning are crucial.

Scheduling for the intelligent manufacturing production line is dynamic, unlike that in job shops, where a relatively rigid schedule is followed to perform a set of operations. The scheduling system for the intelligent manufacturing production line specifies the types of the operations to be

Fig. 13-1-1 Intelligent manufacturing production line

performed on each part and it identifies the machines or manufacturing cells to be used. Dynamic scheduling is capable of responding to the quick changes in the product type and so is responsive to the real-time decisions.[3]

Because of the flexibility in the intelligent manufacturing production line, no setup time is wasted in switching between manufacturing operations and the system is capable of performing different operations in different orders and on different machines. However, the characteristics, performance and reliability of each unit in the system must be checked to ensure that parts moving from workstation to workstation are of acceptable quality and dimensional accuracy.[4]

3. Economic justification of the intelligent manufacturing production line

Intelligent manufacturing production line installations are very capital-intensive, typically starting at well over \$1 million. Consequently, a thorough cost-benefit analysis must be conducted before a final decision is made.

This analysis should include such factors as the cost of capital, of energy, of materials, of labor, the expected markets for the products to be manufactured and any anticipated fluctuations in the market demand and product type.[5] An additional factor is the time and effort required for the installing and debugging the system.

Typically, an intelligent manufacturing production line system can take 2-5 years to install and at least six months to debug. Although the intelligent manufacturing production line requires few, if any, machine operators, the personnel in charge of the total operation must be trained and highly skilled. These personnel include manufacturing engineers, computer programmers and maintenance engineers.

Compared to the conventional manufacturing system, some benefits of the intelligent manufacturing production line are the following.

(1) Parts can be produced randomly in batch sizes as small as one and at a lower unit cost.

(2) Direct labor and inventories are reduced to yield the major saving over the conventional systems.

(3) The lead times required for the product changes are shorter.

(4) Production is more reliable, because the system is self-correcting and the product quality is uniform.

Section Ⅱ New Words and Phrases

intelligent manufacturing production line	智能制造生产线
CNC (computer numerical control)	计算机数控
integrate	*vt*. 集成，使一体化，积分 *v*. 结合
material-handling	物料输送，原材料处理
successive	*adj*. 继承的，连续的
workstation	*n*. 工作站
optimize	*vt*. 使最优化
grinding	*n*. 磨削
cutting	*n*. 切削
powder metallurgy	粉末冶金
heat treating	热处理
finishing	*n*. 带式磨光，饰面，表面修饰，擦光
raw materials	原材料
inspection	*n*. 检查，检验，视察
assembly	*n*. 集合，装配，集会，集结，汇编
machine-tool	母机，机床
sophistication	*n*. 老练，成熟，精致，世故
configuration	*n*. 构造，结构，配置，外形
incorporate	*adj*. 合并的，结社的 *v*. 合并，组成公司
yield	*n*. 产量，收益 *v*. 生成 *vi*. (~to) 屈服，屈从
milling machine	铣床
drill press	钻床，立式钻床
grinder	磨床，磨机，研磨机，磨工
punching	凿孔，冲板，冲压，冲孔
shearing	*n*. 剪切
furnace	*n*. 炉子，熔炉
forging machine	锻造机
trimming press	冲拔罐修边机
retrieval system	*n*. 回收系统
transfer mechanism	传输机械装置
pallet	*n*. 托盘，货盘
terminal	*n*. 终点站，终端，接线端
idle	*adj*. 空闲的 *vi*. (机器) 空转，闲置
schedule	动态调度，工作调度

capital-intensive 资金集约型
debug *v.* 调试

Section Ⅲ Notes to Complex Sentences

[1] These steps may involve one or more processes and operations (such as machining, grinding, cutting, forming, powder metallurgy, heat treating, and finishing) as well as handling of the raw materials, inspection and assembly.

这些步骤可能包括一个或多个程序和操作（比如加工、磨削、切削、成形、粉末冶金、热处理和修整），还有处理原材料、检查和汇编。

as well as…意为“也，还”，常用来连接两个并列的成分，它强调的是前一项，后一项只是顺便提及。

The most common applications of the intelligent manufacturing production line to date have been in the machining and assembly operations.

到目前为止，智能生产线最常见的应用是机加工和装配操作。

to date…至今，到目前为止。

[2] The system is capable of the transporting raw materials, blanks, and parts in various stages of the completion to any machine (in random order) and at any time.

这个系统能够在任何时间向任何机器（随机顺序）传输不同阶段加工完成的原材料、毛坯及零件。

[3] Dynamic scheduling is capable of responding to the quick changes in product type and so is responsive to the real-time decisions.

动态调度对产品类型的快速变化能够做出反应，因此能对实时决策做出响应。

[4] However, the characteristics, performance and reliability of each unit in the system must be checked to ensure that parts moving from workstation to workstation are of the acceptable quality and dimensional accuracy.

然而，必须对系统中每个制造单元的特点、性能和可靠性进行检验，以确保在工作站之间流动的工件满足质量及尺寸精度方面的要求。

that parts moving from…是宾语从句，作 ensure 的宾语。

[5] This analysis should include such factors as the cost of the capital, energy, materials, labor and the expected markets for the products to be manufactured, and any anticipated fluctuations in the market demand and product type.

这个分析应该包括这些因素，如资金、能源、材料和劳动力的成本、即将投产产品的市场预期以及在市场需求和产品类型方面可能出现的波动。

Section Ⅳ Exercises

Translate the following sentences into Chinese.

(1) Intelligent manufacturing production line represents the highest level of the efficiency,

sophistication and productivity that has been achieved in the manufacturing plants.

(2) For the machining operations, they usually consist of a variety of 3-5 axis machining centers, CNC lathes, milling machines, drill presses and grinders.

(3) Because the intelligent manufacturing production line involves a major capital investment, the efficient machine utilization is essential, that is to say, the machines must not stand idle. Consequently, the proper scheduling and process planning are crucial.

Section V Supplementary Reading

Flexible Manufacturing System

A flexible manufacturing system is a manufacturing system in which there is some amount of flexibility that allows the system to react in the case of changes, whether predicted or unpredicted. This flexibility is generally considered to fall into two categories, which both contain numerous subcategories.

The first category, machine flexibility, covers the system's ability to be changed to produce the new product types and the ability to change the order of the operations executed on a part. The second category is called the routing flexibility, which consists of the ability to use multiple machines to perform the same operation on a part as well as the system's ability to absorb the large-scale changes, such as in the volume, capacity or capability.

Most FMS systems comprise of three main systems. The work machines which are often automated CNC machines are connected by a material handling system to optimize the parts flow and the central control computer which controls material movements and machine flow.

The main advantages of an FMS is its high flexibility in the managing manufacturing resources like time and effort in order to manufacture a new product. The best application of an FMS is found in the production of small sets of products like those from a mass production.

As a summary of the above, listed below are the advantages and disadvantages of FMS:

Preparation time for new products is shorter due to flexibility;

Improved production quality;

Saved labour cost;

Productivity increment;

Saved labour costs must be weighed against the initial cost of FMS;

Drawback of the increased flexibility may decrease the productivity.

Exercises

1. Write T(True) or F(False) beside the following statements about the text.

(1) ________An intelligent manufacturing production line includes all major elements of

the manufacturing to a manual system.

(2) __________The most common applications of the intelligent manufacturing production line to date have been in the assembly operations.

(3) __________The flexibility of the intelligent manufacturing production line is such that it can handle a variety of part configurations and produce them in any order.

(4) __________Material handling is controlled by a distributed computer and performed by manually-controlled vehicles, conveyors and various production mechanisms.

(5) __________The workstations are arranged to yield the greatest efficiency in production, with an orderly flow of the materials, parts and products through the system.

(6) __________The computer control system of the intelligent manufacturing production line is its brain and includes various software and hardware.

(7) __________Consequently, a thorough cost-benefit analysis must be conducted before a final decision is made.

(8) __________Dynamic scheduling is not far capable enough of responding to the changes in the production and so is responsive to the simulated decisions.

2. Choose the best answer.

(1) The scheduling system for the intelligent manufacturing__________ specifies the types of the operations to be performed on each part and it identifies the machines or manufacturing cells to be used.

A. assembly line B. production line

C. assembly method D. production method

(2) The workstations are arranged to__________the greatest efficiency in production, with an orderly flow of the materials, parts and products through the system.

A. output B. create C. yield D. locate

(3) However, the characteristics, performance and reliability of each unit in the system must be checked to ensure that parts moving from workstation to workstation are of acceptable quality and dimensional__________.

A. exactness B. correction C. rightness D. accuracy

(4) Consequently, a thorough__________analysis must be conducted before a final decision is made.

A. cost-benefit B. cost-relation C. cost-spending D. cost-consumption

(5) The best application of an FMS is found in the production of small sets of products like those from a__________.

A. mass manufacturing B. mass production

C. mass making D. mass output

3. Fill in the blanks with words or expressions according to the text.

(1) Through the workstation, different computer __________for the manufacturing process

can be downloaded for each __________ part.

(2) The most common applications of the__________ line to date have been in the __________.

(3) The__________of the intelligent manufacturing production line is such that it can handle a variety of part__________ and produce them in any order.

(4) The system is capable of transporting__________, __________in various stages of the completion to any machine (in random order) and at any time.

(5) This analysis should include such factors as the cost of capital, energy, materials, labor, the __________ for the products to be manufactured and any __________in the market demand and product type.

(6) The first__________, __________, covers the system's ability to be changed to produce the new product types and the ability to change the order of the operations executed on a part.

4. Answer the following questions according to the text.

(1) 什么是智能制造生产线？智能制造生产线单元都有哪些？

(2) 如何安排智能制造生产线的调度？

(3) 柔性制造系统的灵活性体现在哪里？柔性制造系统有什么优缺点？

5. Translate the following paragraphs into Chinese.

(1) Intelligent manufacturing production line systems represent the highest level of the efficiency, sophistication and productivity that has been achieved in the manufacturing plants. The flexibility of the intelligent manufacturing production line is such that it can handle a variety of part configurations and produce them in any order.

(2) Intelligent manufacturing production line systems represent the highest level of the efficiency, sophistication and productivity that has been achieved in the manufacturing plants. The flexibility of the intelligent manufacturing production line is such that it can handle a variety of part configurations and produce them in any order.

(3) This analysis should include such factors as the cost of capital, energy, materials, labor, the expected markets for the products to be manufactured and any anticipated fluctuations in the market demand and product type. An additional factor is the time and effort required for the installing and debugging the system.

单元评价

通过本单元课文的阅读，单词、短语练习，扩展阅读及习题练习，使学生能够掌握智能生产线及其应用相关的词汇和表达语句，能够自行完成相关英文的翻译，具备与技术人员利用英语进行智能生产线相关内容交流的能力，同时通过本单元的学习，使学生能够树立正确分析问题、解决问题的思想，激发创新思维，并形成科学严谨、一丝不苟的工作作风。

单元 14

机电设备保养与维护

【学习目标】

知识目标：

1. 熟悉机电设备保养与维护的相关英语词汇。

2. 掌握与技术人员交流机电设备保养与维护时常用的英语表达语句。

技能目标：

1. 具备机电设备保养与维护相关英语词汇跟读技巧，能够运用英语熟练翻译机电设备保养流程及相关知识。

2. 具备使用英语描述机电设备保养流程及注意事项的能力。

素养目标：

1. 遵守职业道德准则和行为规范。

2. 形成着装整洁、文明生产等良好的职业习惯与职业态度。

Section I Texts

1. Concept of Electromechanical Equipment Maintenance

The premise and basis for the use of mechanical and electrical equipment is the daily maintenance of the equipment. The maintenance of the equipment covers a wide range, including cleaning, inspection, lubrication, fastening and adjustment and other daily maintenance work to prevent the equipment from the deterioration and maintain the equipment performance; the necessary inspection to determine the degree of the equipment deterioration, the performance reduction; and the repair activities to restore the equipment performance.[1]

2. Contents of Electromechanical Equipment Maintenance

The maintenance of the equipment is mainly based on the use requirements of different kinds of equipment, according to the instruction manual for the targeted maintenance. It mainly includes rust and corrosion prevention, cleaning, oil change, power supply maintenance and other necessary adjustment work, such as adjusting the fit clearance, fastening parts, and so on.

Rust and corrosion prevention are the most complex work in the equipment maintenance. In manufacturing, the appropriate physical or electrochemical rust prevention methods should be selected to protect the whole equipment and parts according to the environment of the equipment, considering the temperature, humidity, industrial atmosphere (such as sulfur dioxide, carbon dioxide and other acid gases), dust and other factors.[2] When maintaining equipment, attention should also be paid to the health of the frame and housing, the surface and internal pollutants of the equipment should be cleaned regularly and the parts easily to wear should be added oil or replaced in time. It is also necessary to check regularly whether all kinds of the protection and protective facilities are in good condition, such as grounding devices, protective barriers, signs, warning signs, etc. In addition, the power supply is the power source of all the mechanical and electrical equipment and the normal operation of the equipment depends on a stable and reliable power supply environment. Therefore, it is necessary to regularly check the power lines, distribution cabinets, distribution boxes and other power facilities in order to strengthen the management of the power supply lines, prevent the occurrence of the rodent damage, aging and other situations and deal with the problems in time.

3. The Importance of Electromechanical Equipment Maintenance

(1) To ensure the normal operation of mechanical and electrical equipment.

Mechanical and electrical equipment is an important guarantee for the product processing of many enterprises in the new era. To do its maintenance work well, the problems and hidden dangers of equipment can be found in time and solved at the first time, so as to lay a good foundation for the safe operation of the equipment and improve the production order of the enterprise products effectively.

Combined a large number of the practices, in the manufacturing and processing of the enterprises, once the failure of the mechanical and electrical equipment will not only affect the working efficiency, but also easily cause a variety of the production accidents.[3] Therefore, it is very important to improve the maintenance of the mechanical and electrical equipment, which is also the key to improve the application rate of the equipment and ensure the normal operation of all the functional modules and parts.

At the present stage, the production scale of the domestic enterprises is constantly expanding and many mechanical and electrical equipment needs to be put into use for a long time, which also increases the internal structure and parts wear rate to a certain extent. Only if doing good maintenance work according to the operation condition, operating environment and other factors of the mechanical and electrical equipment, can we judge whether the equipment is in normal operation state timely and accurately. It is beneficial to ensure the safe usage and normal operation of the equipment.[4]

(2) Stabilized mechanical and electrical equipment functionality.

Doing a good job in mechanical and electrical equipment repair and maintenance work play an active role in stabilizing the equipment functions. As we all know, after mechanical and electrical equipment is put into the utilization, affected by operating environment and other factors, it is easy to cause some functions to be restricted, which requires the maintenance department to conduct in-depth analysis of the mechanical and electrical equipment itself and operating environment, sum up various factors affecting the function of the equipment and take effective measures to solve the problems.

In the manufacturing of the enterprises, once the function of the mechanical and electrical equipment does not conform to the standard, it is mostly caused by the unreasonable operation parameters of the equipment and the disorder of the operation procedure, which needs to be adjusted through maintenance in time.[5] Especially under the new situation, the processing precision of the enterprise products is getting higher and higher. If there are obstacles in the function of the mechanical and electrical equipment, it will lead to the deviation between the products produced and the precision requirements, which is not conducive to improving the qualified rate of the products, causing huge economic losses for the enterprises. Regular maintenance of the mechanical and electrical equipment can effectively improve the stability and reliability of the equipment functions, thus ensure the orderly progress of the production tasks.[6]

4. Schedule Maintenance Precautions of Electromechanical Equipment

In order to improve the operation stability of the mechanical and electrical equipment, eliminate the potential faults as far as possible and provide guarantee for the manufacturing and processing of the enterprises, we need pay attention to the following items in daily maintenance.

First, do a good job of the mechanical and electrical equipment daily cleaning work, ensure that the equipment inside and outside clean and tidy, avoid the being excessive dust or debris blockage overheating, stop running and other failures. In addition, in the daily maintenance process, we also need pay attention to observing whether the oil hole has the sliding surface oil leakage problem or not, if there is to find the cause and solve in time. Before the equipment is put into usage, attention should be paid to check the surface integrity of the equipment, and do a good job in the test to ensure that the stable operation can be used. If there is a problem of air leakage and oil leakage, communicate with the manufacturer and maintenance timely.

Second, after the mechanical and electrical equipment is put into the usage, it is necessary to set it in a proper position and the lines, pipelines, etc., should be reasonably planned to ensure that the equipment is always in good condition.

Third, always observe the lubrication of the parts of the mechanical and electrical equipment. If the amount of the oil is insufficient, it should be supplemented in time to avoid the phenomenon of friction between parts due to the lack of the oil. On this basis, we should also pay attention to the measurement of the oil pressure, choose quality products when filling the oil and clean the oil gun and oil cup regularly.

Fourth, the technicians should master the operation methods of the mechanical and electrical equipment, they should strictly regulate their own behavior. In the process of the usage, they should pay attention to the use of the appropriate safety protection means to ensure the stable operation of the equipment and lay the foundation for the orderly production.

Section Ⅱ New Words and Phrases

electromechanical equipment	机电设备
maintenance	*n.* 维护，保养；保持，维持
premise	*n.* 前提，假设；*v.* 以……为基础，以……为前提
basis	*n.* 基础，要素；基准，方式；理由，根据
range	*n.* 范围，界限；*v.* 变化，变动
cleaning	*n.* 清洗；清除；去污
lubrication	*n.* 润滑；润滑作用
fasten	*v.* 紧固，扣紧
adjustment	*n.* 调整，校正
performance	*n.* 工作情况，表现；性能；*adj.* 高性能的
degree	*n.* 度，度数（温度单位）；（角的）度，度数；程度
deterioration	*n.* 恶化，设备劣化
reduction	*n.* 减小，降低；性能降低
repair	*vt.* 修理，修补；*n.* 修理，补救
restore	*vt.* 恢复；使复原，使复位；修复，整修
rust and corrosion prevention	防锈、防腐蚀
oil change	更换机油，更换润滑油
power supply maintenance	电源维护
fit clearance	配合间隙
complex	*adj.* 复杂的；组合的，合成的
physical or electrochemical rust prevention	物理或电化学防锈
temperature	*n.* 温度，气温
humidity	*n.* 潮湿，湿气；湿度
atmosphere	*n.* 大气，大气层；空气；气氛，环境
sulfur dioxide	二氧化硫
carbon dioxide	二氧化碳
acid gas	*n.* 酸性气体
dust	*n.* 沙尘；灰尘；粉末；尘雾
housing	*n.*（机器的）外壳，外罩
pollutant	*n.* 污染物

wear	*v.* 销蚀，磨损；穿（衣服），戴（首饰等）
facility	*n.* 设施，设备；天赋，才能；（机器等的）特别装置
grounding device	接地装置
protective barrier	遮壁；防护栏杆；防护栅栏
warning sign	警告标志，警示牌
depends on	取决于；依赖于
stable	*adj.* 稳定的，牢固的；稳重的，沉稳的
reliable	*adj.* 可靠的，可信赖的；耐用的，性能稳定的
power lines	电力线，输电线
distribution box	配电箱；配电盒
hidden danger	安全隐患
effectively	*adv.* 有效地；实际上
practice	*n.* 实践；习惯做法；*v.* 练习；遵循……做法
key	*n.* 关键；钥匙；（电脑、打字机或乐器的）键
at present stage	现阶段；目前
scale	*n.* 规模，范围；标尺，刻度尺；*adj.* 按比例缩小的
domestic	*adj.* 国内的，本国的；家用的，家庭的
timely	*adv.* 及时地
accurately	*adv.* 精确地，准确地
beneficial	*adj.* 有益的，有利的
stabilized	*adj.* 稳定的；*v.* 稳定（stabilize 的过去分词）
role	*n.* 角色；作用，职责
restricted	*adj.* 有限的，受限制的；*v.* 约束，限制（行动或活动）
conform	*vi.* 遵守，符合；一致，相吻合
standard	*n.* 标准，水平，规范；*adj.* 普通的，标准的
potential fault	故障隐患
precaution	*n.* 防范；预防措施；预警

Section Ⅲ Notes to Complex Sentences

[1] The maintenance of the equipment covers a wide range, including cleaning, inspection, lubrication, fastening and adjustment and other daily maintenance work to prevent the equipment

from the deterioration and maintain the equipment performance, necessary inspection to determine the degree of equipment deterioration, performance reduction, repair activities to restore the equipment performance.

设备维护保养包含的范围较广，包括：为防止设备劣化，维持设备性能而进行的清扫、检查、润滑、紧固以及调整等日常维护保养工作；为测定设备劣化程度或性能降低程度而进行的必要检查；为恢复设备性能而进行的修理活动。

[2] Rust and corrosion prevention is the most complex work in the equipment maintenance. In manufacturing, the appropriate physical or electrochemical rust prevention methods should be selected to protect the whole equipment and parts according to the environment of the equipment, considering the temperature, humidity, industrial atmosphere (such as sulfur dioxide, carbon dioxide and other acid gases), dust and other factors.

防锈和防蚀是设备保养中最为复杂的工作。在生产中，应该根据设备所处的环境综合考虑温度、湿度、工业性气氛（如二氧化硫、二氧化碳等酸性气体）、粉尘等因素，选择恰当的物理防锈或电化学防锈方法对设备整体和零件加以保护。

[3] Combined a large number of practices, in the manufacturing and processing of enterprises, once the failure of mechanical and electrical equipment occur, it will not only affect the working efficiency, but also easily cause a variety of the production accidents.

结合大量实践来看，在企业生产加工中，一旦机电设备产生故障，不仅会影响作业效率，还容易引发各种生产事故。

[4] Only if doing good maintenance work according to the operation condition, operating environment and other factors of the mechanical and electrical equipment, can we judge whether the equipment is in normal operation state timely and accurately.

只有根据机电设备运行情况、运行环境等要素做好维修保养工作，才能及时、准确地判断设备是否处于正常运行状态。

[5] In the manufacturing of the enterprises, once the function of the mechanical and electrical equipment does not conform to the standard, it is mostly caused by the unreasonable operation parameters of the equipment and the disorder of the operation procedure, which needs to be adjusted through maintenance in time.

在企业生产中，一旦机电设备出现功能与标准不符的情况，大多由设备运行参数不合理、运行程序紊乱引起，需要及时通过维修保养进行调整。

[6] Regular maintenance of the mechanical and electrical equipment can effectively improve the stability and reliability of the equipment functions, thus ensure the orderly progress of the production tasks.

定期开展机电设备维修、保养工作，可以有效地提高设备功能的稳定性和可靠性，从而保证生产任务有条不紊地进行。

Section Ⅳ Exercises

Translate the following text into Chinese.

Daily Maintenance Level of Electromechanical Equipment

During the operation of the mechanical and electrical equipment, in order to avoid the potential faults as far as possible, it is necessary to divide the daily maintenance levels. Common levels are Level 1 and Level 2 maintenance, which are as follows.

(1) Level 1 maintenance.

This maintenance method is mainly based on the manual inspection of the running status of the mechanical and electrical equipment, we should determine whether the equipment has hidden trouble by analyzing the results and disassemble and clean the parts in the equipment, but ensure that the cleaning method is professional, such as: cleaning of the oil transmission line need to use hair felt and other special tools, if the inspection process found that the parts are not necessary for the maintenance or exceed the service cycle, it must be replaced in time.

After the parts are cleaned and replaced, the parts shall be assembled according to the standard process to ensure that the distance between all parts is reasonable and coordinated. Under the normal circumstances, the Level 1 maintenance of the mechanical and electrical equipment needs to last for about 5 h. After completing the maintenance, the processing should be recorded in detail, including the inspection method and inspection time.

(2) Level 2 maintenance.

This maintenance mode requires the cooperation of the operators and maintenance personnel of the mechanical and electrical equipment and complete the equipment maintenance work together. Specifically, it is necessary to disassemble the energy-consuming parts of the mechanical and electrical equipment first and then check whether there are problems such as wear and tear of the parts and, at the same time, clean up the dirt on the parts and replenish the oil.

Section Ⅴ Supplementary Reading

Specific Measures for Electromechanical Equipment Repair and Maintenance

1. Master Repair Methods of the Mechanical and Electrical Equipment

Firstly, if the mechanical and electrical equipment fails during the operation, the maintenance personnel must cut off the power supply at the first time and then deal with the fault by replacing the parts, etc., to ensure that the equipment quickly returns to normal. If the fault type of the mechanical and electrical equipment is complex and the damage degree is large, the maintenance department needs to carry out extensive maintenance.

In the specific implementation process, the maintenance personnel should operate in strict accordance with the current norms and relevant systems and unavoidably disassemble the equipment blindly, which is also the key to ensure the maintenance effect. In addition, the electromechanical equipment introduced by the current enterprise generally has the characteristics of the automation, intelligence and integration and its components assembly and internal structure are more complex, so the faults generated in the operation are also commonly hidden, which not only increases the difficulty of the maintenance, but also consumes a lot of maintenance costs. In order to avoid the potential faults as much as possible, the technical personnel are required to do maintenance work regularly, check the running status of the parts one by one and ensure that the equipment is in stable operation. This requires technical personnel to master the equipment maintenance methods, who clean, adjust and reinforce the equipment through daily and regular maintenance and reduce the incidence of the hidden accidents from the root.

The maintenance level distinction mentioned above is actually the distinction of the equipment and time, which will have a huge impact on the running state of the equipment. Usually, the higher the maintenance level of electromechanical equipment, the larger the maintenance range will be. Although many equipment can be restored to the normal state through the maintenance after the failure, it cannot be guaranteed that they are intact.

2. Master Maintenance Methods of Mechanical and Electrical Equipment

During the normal operation of the mechanical and electrical equipment, it is greatly affected by the environmental temperature and humidity and other factors. If the operation environment is harsh or the operation time is relatively long, it is necessary for the staff to increase the maintenance efforts, appropriately increase the maintenance frequency, take reasonable methods to do equipment management and create a good environment for the operation of the mechanical and electrical equipment by improving the operating environment.

In the specific implementation process, in order to achieve the ideal maintenance effect as far as possible, it is necessary for the technical personnel to establish the advanced consciousness, improve the innovation ability, constantly update the maintenance methods of the mechanical and electrical equipment and effectively integrate traditional manual maintenance with the advanced science and technology. To be specific, the maintenance department needs to combine the operating conditions, operating environment, service life and other factors of the different mechanical and electrical equipment to develop the maintenance plans in line with them and organize a team of the professional talents to complete the maintenance work to ensure that the maintenance system is in place.

Exercises

1. Write T(True) or F(False) beside the following statements about the text.

(1) __________The premise and basis for the use of mechanical and electrical equipment is

the daily maintenance of the equipment.

(2) __________The maintenance of the equipment is only based on the use requirements of all equipment, according to the instruction manual for the targeted maintenance.

(3) __________ Mechanical and electrical equipment is an important guarantee for the product processing of many enterprises in the new era.

(4) __________Doing a good job in mechanical and electrical equipment repair plays an active role in the utilization of the equipment functions.

(5) __________Especially in the past situation, the processing decision of the enterprise products is getting harder and harder.

(6) __________ Regular maintenance of the mechanical and electrical equipment can effectively improve the stability and reliability of the equipment functions, thus ensure the orderly progress of the production tasks.

(7) __________ If the fault type of the mechanical and electrical equipment is simple and the damage degree is not so large, the maintenance department does not need to carry out extensive maintenance.

(8) __________After completing the maintenance, the processing may be recorded in detail, including the inspection and maintenance method and time.

2. Choose the best answer.

(1) It mainly includes rust and corrosion prevention, cleaning, oil change, power supply maintenance and other necessary__________work, such as adjusting the fit clearance, fastening parts and so on.

A. repair B. maintenance C. management D. adjustment

(2) Mechanical and electrical equipment is an important __________ for the product processing of many enterprises in the new era.

A. safety B. security C. guarantee D. safeguard

(3) Regular maintenance of the mechanical and electrical equipment can effectively improve the__________of the equipment functions, thus ensure the orderly progress of the production tasks.

A. continuity and stability B. continuity and reliability

C. stability and reliability D. reliability and trust

(4) Although many equipment can be restored to the normal state through the maintenance after the failure, it cannot be guaranteed that they are__________.

A. intact B. contact C. unengaged D. engaged

(5) During the normal__________of the mechanical and electrical equipment, it is greatly affected by the environmental temperature and humidity and other factors.

A. conduction B. implementation C. experiment D. operation

3. Fill in the blanks with words according to the text.

(1) __________and__________are the most complex work in the equipment maintenance.

(2) Regular maintenance of the __________ can effectively improve the __________ of the equipment functions, thus ensure the orderly progress of the production tasks.

(3) Therefore, it is necessary to regularly check the power lines, distribution cabinets, distribution boxes and other power facilities in order to strengthen the management of the __________, prevent the __________ of the rodent damage, aging and other situations and deal with the problems in time.

(4) On this basis, we should also pay attention to the __________ of the oil pressure, choose __________ when filling the oil and clean the oil gun and oil cup regularly.

(5) Before the equipment is put into usage, attention should be paid to check the __________ of the equipment, and do a good job in the test to ensure that the __________ can be used.

(6) This __________ requires the cooperation of the operators and maintenance personnel of the mechanical and electrical equipment and complete the __________ work together.

4. Answer the following questions according to the text.

(1) 机电设备维修与保养的概念是什么？机电设备维修与保养工作的内容主要是什么？

(2) 机电设备维修与保养的重要性是如何体现的？

(3) 机电设备维修与保养的方法是什么？

5. Translate the following paragraphs into Chinese.

(1) The maintenance of the equipment covers a wide range, including cleaning, inspection, lubrication, fastening and adjustment and other daily maintenance work to prevent the equipment from the deterioration and maintain the equipment performance for the necessary inspection to determine the degree of the equipment deterioration, the performance reduction and the repair activities to restore the equipment performance.

(2) Mechanical and electrical equipment is an important guarantee for the product processing of many enterprises in the new era. To do its maintenance work well, the problems and hidden dangers of equipment can be found in time and solved at the first time, so as to lay a good foundation for the safe operation of the equipment and improve the production order of the enterprise products effectively.

(3) The maintenance level distinction mentioned above is actually the distinction of the equipment and time, which will have a huge impact on the running state of the equipment. Usually, the higher the maintenance level of electromechanical equipment, the larger the maintenance range will be. Although many equipment can be restored to the normal state through the maintenance after the failure, it cannot be guaranteed that they are intact.

单元评价

通过本单元课文的阅读，单词、短语练习，扩展阅读及习题练习，使学生能够掌握机电设备保养与维护相关的英语词汇和表达语句，能够自行完成相关英文的翻译，并具备与技术人员利用英语进行电设备保养与维护相关内容交流的能力，同时，通过本单元的学习，使学生能够遵守职业道德准则和行为规范，形成着装整洁、文明生产等良好的职业习惯与职业态度。

单元 15

生产现场优化与数字化管理

【学习目标】

知识目标：

1. 熟悉生产现场优化与数字化管理相关英语词汇。
2. 掌握与技术人员交流生产现场优化与数字化管理时常用的英语表达语句。

技能目标：

1. 具备生产现场优化与数字化管理相关英语词汇跟读技巧，能够熟练翻译生产现场优化与数字化管理英语资料。
2. 具备使用英语描述生产现场优化与数字化管理的能力。

素养目标：

1. 获得较强的语言、文字表达和社会沟通能力。
2. 形成不断进取、求实创新和超越自我的精神。
3. 获得善于发现问题并从根本上解决问题的能力。

Section I Texts

Production Site Management

Production site management is an important part of the enterprise management, and the management level directly affects the production efficiency and economic benefits of the enterprises.[1] Optimizing production site management is the premise and foundation of the enterprise management to achieve the overall optimization. The survival and development of the enterprises follow the objective law of the survival of the fittest, which mainly depends on the market products and two factors of the management level. The important symbols of an enterprise's competitive advantages include the product quality, service level, delivery time and production cost. Stable product quality is the key to the success for an enterprise in the fierce market competition.[2] Scientific and meticulous management of the production site is an important guarantee of the high quality and low price.

1. Introduction

Production site management optimization aims to improve the management level of the enterprise, improve the awareness and skill level of the relevant personnel on the production site, take how to improve the production mobility as the main line and improve the production efficiency, quality, delivery time, safety, environmental protection, staff morale and so on. Through the training of the production site management contents, employees can change their working concept and improve their working skills, so that the employees can actively participate in the site management.[3] While employees themselves are improved in various aspects, enterprises can get higher returns.

2. Significance of the Production Site Management

(1) Establish enterprise advantages.

Improving the level of the production site management is the foundation of establishing the enterprise superiority strengths. The strengths of an enterprise refers to the strengths that an enterprise has which is higher than its competitors in competing for the market and customers. The advantages of the enterprises include capital, manpower, technology, marketing, product quality, product cost and so on. The establishment of a comprehensive management system centered on the production site is a sharp tool to tap production potential and improve the economic benefits.

(2) Comprehensively improve the quality of the enterprises.

Improving the level of the production site management is the premise and guarantee to improve the quality of the enterprise management, production technology, basic work, staff quality and spiritual civilization quality, otherwise the overall improvement of the enterprise quality can only be hollow words. The quality of the site management is an important sign to measure the level of the enterprise management.

(3) Improve the economic benefits of the enterprises.

The improvement of the economic benefits of the enterprises must be grabbed from the production site, because the enterprise benefits from the labor of all the workers who created the value and these values are mostly realized in the production site. Only through strengthening the production site management, with the lowest input to obtain the maximum output, can the rational allocation and effective use of the production factors be realized.

Digital Management

Digital production management refers to the management activities and methods that use computer, communication, network and other technologies to quantify the management objects and management behaviors through the statistical techniques to achieve the research and development, planning, organization, production, coordination, sales, service, innovation and other functions.[4]

Contents of the digital production management are listed below.

(1) Lean operation improvement—optimize and improve the enterprise operation and production site with lean thinking.

(2) Intelligent management on the site—digital intelligent management and control on the

production site through the industrial Internet of Things.

(3) Equipment operation and maintenance management—grasp the real-time situation of the equipment through the computer networking platform to realize the remote supervision.

(4) Quality monitoring and management—control important links of the production to realize the traceability management of the whole process.

(5) Warehouse logistics management—scanning and induction are adopted for data acquisition to realize the intelligent warehouse operations.

Section Ⅱ New Words and Phrases

production site management	生产现场管理
enterprise	*n.* 企业，事业单位；事业；创业，企业经营
optimization	*n.* 最佳化，优化
survival	*n.* 继续生存，存活，幸存；*adj.* 幸存的，赖以生存的
law	*n.* 法律（体系）；法令，法规；规律，法则，定律
factor	*n.* 因素（factor 的复数）*v.* 把……作为因素考虑
scientific	*adj.* 科学的，与科学有关的；细致严谨的
meticulous	*adj.* 一丝不苟的，注意细节的
awareness	*n.* 认识，意识；感悟能力
relevant	*adj.* 有关的，切题的；正确的，适宜的
mobility	*n.* 流动能力；活动性，灵活性
employee	*n.* 员工；雇员；从业人员
aspect	*n.* 方面，特色；朝向，方位；外表，外观
foundation	*n.* 地基，基础；基本原理，根据；基金会
superiority	*n.* 优越，优势；优越感，骄傲自大
capital	*n.* 资本，资金；资本家，资方；首都，首府，省会
manpower	*n.* 人力；人力资源；劳动力
comprehensively	*adv.* 包括一切地；完全地；彻底地；全面地
premise	*n.* 前提，假设；*v.* 以……为基础，以……为前提
guarantee	*v.* 确保，保证；*n.* 保证，担保；保修单，质量保证书
enterprise management	企业管理
sign	*n.* 指示牌，标志；迹象，征兆；符号，记号；*v.* 签名，署名
grab	*v.* 攫取，抓住；利用，抓住（机会）；*n.* 攫取，赚取
obtain	*v.* 得到，获得；存在，通用
digital management	数字化管理
organization	*n.* 组织，机构；组织工作，筹备工作

coordination	*n.* 协调，配合；身体的协调性；配位
service	*n.* 服务；接待，服务；*v.* 维修，检修
innovation	*n.* 新事物，新方法；革新，创新
lean	*v.*（身体）倾斜；倚靠，靠在；*adj.* 精简的，效率高的
monitoring	*n.* 监视，监控；检验，检查；*v.* 监视，监听，监督
warehouse	*n.* 仓库，货栈，货仓；*v.* 将（商品）存库，使入库

Section Ⅲ Notes to Complex Sentences

[1] Production site management is an important part of the enterprise management and the management level directly affects the production efficiency and economic benefits of the enterprises.

生产现场管理是企业管理的重要组成部分，管理水平直接影响企业的生产效率和经济效益。

[2] The important symbols of an enterprise's competitive advantages include the product quality, service level, delivery time and production cost. Stable product quality is the key to the success for an enterprise in the fierce market competition.

企业竞争优势的重要标志包括产品质量、服务水平、交货期以及生产成本。稳定的产品质量是企业在激烈的市场竞争中取胜的法宝。

[3] Through the training of the production site management contents, employees can change their working concept and improve their working skills, so that the employees can actively participate in the site management.

通过生产现场管理内容的培训可以改变员工的工作理念、提高工作技能，使员工能积极参与到现场管理中去。

[4] Digital production management refers to the management activities and methods that use computer, communication, network and other technologies to quantify the management objects and management behaviors...

数字化生产管理是指利用计算机、通信、网络等技术，量化管理对象与管理行为的管理活动和方法……

Section Ⅳ Exercises

Translate the following paragraphs into Chinese.

Digital production management system has become an indispensable part of the modern manufacturing industry and its main role is to enhance the enterprise's control over the production process, optimize the production process and improve the production efficiency and quality.

The features of the digital production management system include real-time monitoring, automatic management, improving production efficiency, improving quality of product and improving production safety, etc. These features can help the enterprises to optimize the production process, improve production efficiency, product quality and enhance the safety of the production site. Therefore, the application of the digital production management system has become an important means to optimize the production process and promote the improvement of the production efficiency in the modern manufacturing industry.

Section V Supplementary Reading

Strategy of Optimizing Production Site Management

There are many methods to optimize the production site management, such as the work study, Kanban management, the group technology, the total control management method, the first-class production mode, the full load working method, the multi-machine tool care method, ABC classification method, the economic order batch method, the equipment point verification and repair, the value engineering and the field process quality control method and the implementation of standardized operations. The establishment of the production line operation process serves as the main body of the labor organization. In recent years, the production site management methods are widely used in various enterprises are: "5S" activity, the location management, the visual management and Kanban management. Most of the above methods have different application scopes and conditions, so they should be chosen correctly when applied.

1. Standardized Management of the Production Site

Production site management is the comprehensive management of the first line of the production, standardized management and abnormal management which complement each other. The enterprise should take the construction of the production site management standards as the goal, check the standards, find the abnormal phenomenon of the production site, analyze the reason and make improvements, which is the whole work of the production site management. The implementation of the standardized management on the production site should accomplish the standardization of the operation procedures, product production equalization, equipment integrity, safety, civilized production institutionalization, site layout targeting, product quality automation and morale encouraging diversification.

2. Implement "5S" Management Activities in the Workshop

"5S" management is the effective management of the production factors such as the materials, machinery, methods and personnel on the production site. "5S" includes arrangement, rectification, cleaning and so on. Enterprises can continuously enrich and develop "5S" management according to their own characteristics. "5S" activity can improve the employee morale, product quality and production efficiency, which is one of the cornerstones for the effective implementation of other management activities. "5S" can also improve the production environment and has played a huge role in establishing a high degree of the enterprise standardization, safety and standardized production, reducing costs, shaping the image of the enterprise, creating a relaxing and enjoyable working environment and site improvement. "5S" management has become an effective method of the management standardization today.

3. Workshop Fixed Management

Location management is based on the scientific location of the goods in the site as the premise, scientific analysis of the relationship between people, goods and places in the production site, research the best combination state of a scientific management method. With the complete information system as the medium, the location management standardizes the production site through the arrangement and rectification of the production site, removes the unnecessary items in the production, optimizes the placement of the needed items and makes them easily accessible, so as to promote the civilized and scientific management of the production site and achieve the purpose of the efficient production, safe production and high-quality production. To realize the effective combination of people and things in the production process is the deepening and development of "5S" activities. Elements of the location management activities in a branch factory (workshop) include the location of the site, each process, station, machine, tool box, warehouse, etc.

4. Workshop Visual Management

Visual management is the basic means of the visual signal display. On the premise of reflecting the requirements and intentions of the managers as much as possible, most potential abnormalities of the factory are displayed and the management is made public and transparent, so as to promote the autonomous management or control. Its features include on-site operators which can visually display their suggestions and impressions and communicate with the leaders, colleagues and workmates. All the elements of the factory workshop are applicable to the visual management, such as the services, products, semi-finished products, raw materials, spare parts, equipment, fixture, mold, measuring tools, handling tools, shelves, channels, places, methods, bills, standards, notices, people, etc.

Exercises

1. Write T(True) or F(False) beside the following statements about the text.

(1) ____________ Production site management is an important part of the enterprise management, and the management level directly affects the production efficiency and economic benefits of the enterprises.

(2) ___________Stable product quality is not the key to the success for a company in the equipment market competition .

(3) ___________The important symbols of an enterprise's competitive advantages include the product quality, service level, delivery time and production cost.

(4) ________ Improving the level of the production site management is not the reason of improving the enterprise competitive strengths.

(5) ________ The quality of the site management is an important sign to measure the level of the enterprise management.

(6) ________ Through strengthening the production site management, with the highest input to obtain the maximum output, can the effective use of the production factors be realized.

(7) ________ The establishment of the production line management serves as the only body of the labor market.

(8) ________ "5S" management is the effective management of the production factors such as the materials, machinery, methods and personnel on the production site.

2. Choose the best answer.

(1) Optimizing production site management is the premise and foundation of the enterprise management to achieve the overall ________.

A. result B. consequence C. optimization D. configuration

(2) Stable ________ is the key to the success for an enterprise in the fierce market competition.

A. product quality B. product output

C. product creation D. product manufacturing

(3) The strengths of an enterprise refer to the ________ that an enterprise has which is higher than its competitors in competing for the market and customers.

A. energy B. power C. contribution D. strengths

(4) These features can help the enterprises ________ the production process, improve production efficiency, product quality and enhance the safety of the production site.

A. configure B. progress C. proceed D. optimize

(5) "5S" activity can improve the employee morale, product quality and production efficiency, which is one of the ________ for the effective implementation of other management activities.

A. anchors B. foundations C. cornerstones D. milestones

3. Fill in the blanks with words or expressions according to the text.

(1) ________ is the key to the success for an enterprise in the ________.

(2) The strengths of an enterprise refer to the strengths that an enterprise has which is higher than its ________ in competing for the ________.

(3) The improvement of the ________ of the enterprises must be grabbed from the production site, because the enterprise benefits from the labor of all the workers who created the value and these values are mostly realized in the ________.

(4) "5S" activity can improve the employee morale, product quality and production efficiency, which is one of the cornerstones for the effective ________ of other ________

activities.

(5) Elements of the __________ activities in a __________ (workshop) include the location of the site, each process, station, machine, tool box, warehouse, etc.

(6) Its features include __________ can __________ their suggestions and impressions and communicate with the leaders, colleagues and workmates.

4. Answer the following questions according to the text.

(1) 什么是生产现场管理？如何优化生产现场管理？

(2) 生产现场管理的意义是什么？如何对生产现场进行数字化管理？

(3) 优化生产现场管理的对策有哪些？什么是"5S"管理？如何进行"5S"管理？

5. Translate the following paragraphs into Chinese.

(1) Improving the level of the production site management is the premise and guarantee to improve the quality of the enterprise management, production technology, basic work, staff quality and spiritual civilization quality, otherwise the overall improvement of the enterprise quality can only be hollow words. The quality of the site management is an important sign to measure the level of the enterprise management.

(2) Digital production management refers to the management activities and methods that use computer, communication, network and other technologies to quantify the management objects and management behaviors through the statistical techniques to achieve the research and development, planning, organization, production, coordination, sales, service, innovation and other functions.

(3) "5S" management is the effective management of the production factors such as the materials, machinery, methods and personnel on the production site. "5S" includes arrangement, rectification, cleaning and so on. Enterprises can continuously enrich and develop "5S" management according to their own characteristics. "5S" activity can improve the employee morale, product quality and production efficiency, which is one of the cornerstones for the effective implementation of other management activities. "5S" can also improve the production environment and has played a huge role in establishing a high degree of the enterprise standardization, safety and standardized production, reducing costs, shaping the image of the enterprise, creating a relaxing and enjoyable working environment and site improvement. "5S" management has become an effective method of the management standardization today.

单元评价

通过本单元课文的阅读，单词、短语练习，扩展阅读及习题练习，使学生能够掌握生产现场优化与数字化管理相关的词汇和表达语句，能够自行完成相关英文的翻译，具备与技术人员利用英语进行生产现场优化与数字化管理相关内容交流的能力。同时，通过本单元的学习，使学生能够获得较强的语言、文字表达和社会沟通能力，并形成不断进取、求实创新和超越自我的精神。

参考译文

单元 1　机械制图与建模

课文 1

工程图是工程设计的基础，工程设计的实质要求图纸作为设计过程中进行交流的手段。所以，图纸在设计与加工制作过程中充当共同协议的作用。

工程制图是设计人员走向成功的一项基本技术。在这项技术里，设计人员应具备的一个重要能力是通过草图表达自己的思路是设计的第一要素，即作出三维几何图，让其他工程师、科学家、技术人员和非技术人员能够看懂，这种能力就是宝贵的财富。另一个非常重要的能力，是能够读懂和理解别人绘制的工程图。

良好的沟通能力是一个工程设计人员成功的关键。图纸交流和语言、文字交流一起构成了设计人员的交流手段。在发达国家，图纸语言在专业上的应用并不逊于语言和文字，因此，有人将图纸称为工程师的语言。

工程制图学的研究包括三个方面：术语、技巧以及方法。

工程制图在工程中负责图纸准则的制定以及设计思想的传递。

工程设计是解决实际问题的一个系统工程，其中工程图学提供可视化的支持、工程分析的基础和设计程序的文档。

绘图几何是由一系列准则组成的，这些准则使一个物体的几何元素能够以轮廓的形式被识别出来。它是将三维图形里的角度、形状、尺寸、空隙以及交叉等特点用二维的曲线表达出来。

计算机绘图是用数字计算机来定义、模拟和演示设备、工艺以及系统，最终用于分析和设计工程问题。

几何模型常常作为一个概念、工艺或者系统的数学形式，更具体地还用作数据库信息。基于计算机的几何模型能够非常方便地表达出物体的轮廓、表面以及实体。

工程制图是一门快速发展的技术，传统的绘图装置，比如丁字尺、圆规以及绘图机已经被计算机硬件和软件设备所取代。我们正处于一个让人兴奋的领域，因为我们可以用键盘代替尺子、三角板和圆规等工具，并且将草图放进数据库保存。

现在的工程师将工程制图看成 CAD 的副产品，因为控制设计-制造单元的是电子数据库，设计过程的所有环节都在经历着根本的改变。新产品的模型可以很快地被绘制出来，并且常常可以直接在计算机上模拟其原型。

现在的学生也将从设计程序的角度学习制图，几何建模技术（即以数学为基础的分析技术）以及立体几何图形的制作将是计算机应用的重点方面。为了厘清设计思路和进行模型分析，首先需要学习手工绘图。学生们将通过草图和计算机两种方式来学习和理解绘图，并且，需要学习许多绘图标准，它们涉及物体的一些主要参数，如尺寸、视图等。

若有三个方向的尺寸，则很容易在计算机上依实际需要以二维或三维的模式将图形绘制出来。创建二维模型，比如 *xy* 平面上的，可以直接用 CAD 软件绘制。二维几何模型提供了圆、矩形等圆柱体常用的截面形状，计算机绘图软件都可以提供这些功能。某些专业的应用，比如尺寸的标注等也常常应用在二维图形中。

三维几何模型涉及线框模型、曲面模型及实体模型的建立。一个线框模型显示了一系列由线连接成模型的过程。而实体模型是一个物体的整体形状，包括边界的和内部的所有节点。根据实体模型，计算机常常可以对其做彻底的模型分析。

计算机绘图已经成为一个非常强大的设计工具，它能够显著地加强设计人员面对复杂问题时的创新能力。

信息技术的革命已经如火如荼，我们工作和生活在电子信息技术的快速发展中，一切都发生着明显的变化。计算机的绘图语言也将继续扮演其设计工程师和其他技术人员交流手段的角色，而且我们看到的图形绘制方法的演变还是让人震惊的。这些变化将促进工业化更好地发展，也将促使技术人员设计出更高质量的产品。这样的快速发展也要求 21 世纪的工程师们必须对绘图的基础以及发展有一个非常深入的理解，从而充实到他们的设计程序当中去。

课文 2

一台计算机，我们如何控制并让它运行呢？当然需要一些计算机软件。CAD 和 CAM 这两个术语是指计算机辅助设计和计算机辅助制造。它们是与数字计算机应用相关的技术，用以完成设计与制造的相关功能。这种技术正朝着设计和制造一体化方向发展，现在认为设计与制造在传统上是明显不同的两个方面。CAD/CAM 是未来的计算机集成工厂的基础。

通常认为，计算机辅助设计 CAD 有助于一个设计的创新、修改、分析以及优化。计算机系统由硬件和软件构成，用以完成特殊用户公司需求的专业设计功能。实现 CAD 技术需要的硬件通常包括计算机、一个或多个图形显示器、键盘等其他设备。CAD 软件包含计算机程序和应用程序，计算机程序实现系统上的绘图，而应用程序是为了便于用户实现工程应用而设计的。如组件的应力-应变分析、机构的动态响应、热转换计算以及零件数控编程等都是应用程序的一些典型代表。因每个用户的生产线类型、制造工艺过程以及顾客市场是不同的，因此各用户之间的应用程序是不一样的。

计算机辅助制造 CAM 是指通过计算机直接或者间接地干预工厂的生产资源来计划、管理和控制制造车间的操作。根据定义可见计算机辅助制造系统的应用分为以下两大类。

（1）计算机监测与控制。最直接的应用通常是将计算机连接到制造过程中来监测和控制生产过程。

（2）制造程序。计算机系统的间接应用：为了核实车间生产操作是否正确，从而采用计算机系统与生产过程间接连接来验证。除了通过计算机直接进行过程监测和控制外，计算机辅助制造还包括计算机在工厂生产运营中支撑作用的间接应用。在这些应用中，没有将计算机与制造过程直接连接。相反，为了有效地管理企业资源，计算机采用离线的方式为企业提供计划、日程安排、预测、指导和信息。计算机与生产制造过程关系的形成可用下面给出的图来象征性地表示。计算机与生产操作间的通信和控制连接是计算机的离线连接，通常需要人类来完成这一连接，因此用虚线表示。然而在应用程序中，当前人们只需要在计算机中提供输入指令或者通过计算机输出指令，可执行所需的动作。

单元 2　工程材料与热处理

1. 引言

普通工程材料通常分为金属与非金属。金属可自然地分为铁与非铁金属。目前使用的重要的铁金属有①铸铁、②可锻铸铁、③钢。

应用于工程设计中的一些重要的非铁金属：①轻金属组，如铝及其合金；②铜基合金，如青铜（Cu-Zn），黄铜（Cu-Sn）；③白色金属组，如镍、银、锌等。

2. 铁金属材料

（1）铸铁。

铸铁是一种铁碳硅合金，性硬而脆，碳含量为 1.7%~3%，碳以自由碳原子或碳化铁的形式存在。

（2）可锻铸铁。

可锻铸铁是一种非常纯的铁，其铁含量高达 99.5%。它是由生铁重新熔化而成的，其中含有少量硅和硫，或者少量磷。其性硬，可冶炼、可延展，容易铸造和焊接，但不能承受冲击，可用来生产链条、起重机吊钩、铁轨及类似产品。

（3）钢。

钢是最重要的工程材料，种类繁多，可以满足不同的需要。钢基本上也是铁碳合金，其中碳含量低于 1.7%，以碳化铁的形式存在，并提供钢的硬度和强度。钢可以分为两大类：纯碳钢和合金钢。

①纯碳钢——纯碳钢的性能取决于碳的百分比。其他合金元素含量通常为 0.5%~1%，如 0.5%的硅、1%的锰等。

②合金钢——这种钢添加了除碳以外的其他足够数量的元素，用以提供期望的性能，比如抗磨损性、抗腐蚀性、电磁特性等。其主要合金元素通常有镍（提高强度和硬度）、铬（硬度和强度）、钨（高温硬度）、钒（剪切强度）、锰（热轧强度及热处理强度）和硅（弹性极限）。

3. 非铁金属

含有除铁以外的元素为主要成分的金属称为非铁金属。在实际应用中非铁金属种类繁多，下面只讨论少数几例。

（1）铝。

铝是从铝矿提炼出来的白色金属。纯铝强度差且质软，但在添加少量铜、锰、硅和镁之后变得坚硬且强度好，而且还具备抗腐蚀、轻质及无毒的特点。

（2）镁。

铝镁合金中，镁的含量占 2%~3%，同时还含有 1.75%的铜。因其质地轻、强度高，广泛用于飞机和汽车零部件。

（3）铜合金。

铜是工业上运用最广的非铁金属。铜质软，可塑、可延展，是热和电的良好导体。以下两种重要的铜合金应用最广。

（4）黄铜（铜锌合金）。

基本上是铜与锌的二元合金，其中锌含量可达50%。黄铜极抗腐蚀，容易加工，因此是很好的轴承材料。

（5）青铜（铜锡合金）。

主要是铜锡合金，其中锡含量为5%～25%。锡提供硬度，但锡也会氧化从而导致脆性。它最初用来铸造枪械，但现在用作锅炉辅机、轴瓦、封盖及类似用途。

4. 非金属

非金属材料在实际工程中也因它们的低成本、多样性及热和电的阻抗性得到广泛应用。通常有多种可供选择的非金属，从设计的角度出发，塑料是最重要的一种。非金属材料是人工合成材料，经加热或不加热可以模铸成各种形状。由于非金属材料良好的抗腐蚀性、尺寸稳定性以及相对廉价，故现在广泛用于各行各业。

单元 3　电子电气应用

电梯的控制主要指对电梯在运行过程中的运行方向、层站召唤、负载信号、楼层显示、安全保护等指令信息进行管理。电梯的用途不同，可以有不同的载荷、不同的速度及不同的驱动方式和控制方式，即使相同用途的电梯，也可采用不同的操纵控制方式。但不论电梯使用何种控制方式，所要达到的目标是相同的，即根据轿厢内指令信号、层站召唤信号而自动进行逻辑判定，决定哪一台电梯接收信号，自动定出电梯的运行方向，并按照指令要求通过电气自动控制系统的控制达到预定的控制目的。

1. 电梯控制系统的分类

从控制系统的实现方法来看，电梯的控制系统经历了继电器电梯控制、可编程控制器（PLC）控制、单片机控制、多微机控制多种形式，这些控制方式代表了不同时期电梯控制系统的主流，这些控制系统在目前的在用电梯中都有应用，并且随着大规模集成电路和计算机技术的发展而逐步推陈出新。

2. 电梯控制功能

从按下候梯厅的呼梯按钮到完成电梯的运行全过程并走出轿厢，包含了电梯的许多功能，如呼梯信号的应答，电梯的起动、运行、停车，电梯的开关门等。在电梯控制系统的设计中，有些功能被设定为标准功能，还有些功能被设定为可选功能，根据用户需要可自行选择。下面就集选电梯常用的控制功能作一介绍。

1）呼梯信号的应答

电梯控制系统设置轿厢内层楼选择信号（即内选信号）和大厅层楼呼叫信号（即外呼信号）。这两种信号一般通称为呼梯信号。

（1）内选信号的应答。

①与轿厢运行方向一致的内选信号被登记后，轿厢将会按顺序来应答。

②轿厢停在某层楼或正向某层楼运行且已减速，则该层楼的内选信号不被登记。

③当一个内选信号被登记时，内选按钮灯或发光二极管点亮并显示。

④内选信号将在门区被消号。

（2）外呼信号的应答。

①与轿厢运行方向一致的外呼信号，轿厢将按顺序逐一应答。

②最后一个反向呼叫信号，也将被应答。

③与轿厢运行方向相反的外呼信号，将予以保留，待轿厢应答完运行方向上的最后一个呼梯信号，再换向逐一按顺序应答反方向上的外呼信号。

④轿厢在所停楼层的外呼信号，如与轿厢运行方向相同者，可使门重新开启；若与轿厢运行方向相反者，将予以保留。其条件是轿厢运行到达某层楼，仍保留运行方向，或者在某层已确立运行方向。

⑤外呼信号在下列情况下被消号。

若外呼信号在减速点前被登记，那么该外呼信号便会在减速点被消号。

若外呼信号在减速点后被登记，那么该外呼信号便会在门区消号。

⑥当乘客于中间楼层按上、下两个呼梯按钮时，轿厢运行到该楼层时将顺向信号消除，保留反向呼叫信号。

⑦在满载（LNS）的情况下，满载指示灯亮。此时，不再应答外呼信号，应予保留。

⑧当外呼信号被登记后，外呼按钮指示灯或发光二极管点亮。

2）基站停靠

（1）轿厢完成最后一个呼梯信号，在一定时间内没有任何呼叫信号时，将返回基站。

（2）轿厢返回基站后，处于待梯状态，轿门不开启。

（3）轿厢在返回基站的过程中，若在运行方向前面有呼叫信号出现，则应答此信号；若在运行方向后面有呼叫信号出现，则电梯将于最近停层站停车，不开门，换向应答呼叫信号。

3）驻停

（1）当接通驻停开关时，电梯进入驻停状态，轿厢返回驻停层站。

（2）轿厢到达驻停层时，所有呼叫信号被消除，不再被登记。

（3）轿厢在驻停层驻停，所有的位置指示器、轿厢灯、风扇均被关闭，不运行灯点亮。

（4）轿厢在检修期间，驻停功能不起作用。

（5）独立服务优先于驻停操作，但轿厢只要到达驻停层，轿厢关门驻停。

（6）防犯罪操作，在有一个内选信号的情况下，将优先于驻停功能，但轿厢到达驻停层将实行驻停。

4）紧急停车

（1）人为急停开关

①急停开关一般设在轿顶、机房及底坑，都是供检修人员操作的开关。急停开关采用非自动复位开关，一旦打开急停开关，必须人工复位，电梯才能恢复正常。电梯检修人员可在维修时按下此开关，使电梯不能运行，确保维修人员绝对安全。

②轿厢在运行过程中，如打开急停开关，轿厢将立即停止，并消除所有呼叫信号。

③在急停期间，轿厢不登记任何信号。

④在急停期间，门将关闭。

（2）故障与急停

①电梯设有一系列安全保护装置，如限速器、安全钳、缓冲器、极限开关、安全窗、断绳保护、电机过热等，以保证电梯的安全运行，从而保护乘客的人身安全。电梯在发生故障时，将引发任一安全装置及开关动作或电梯安全回路上的任一节点断开，均将导致电梯急停。

②电梯急停后，消除所有呼叫信号。

③电梯急停后轿厢照明及风扇将不关断。

5）独立服务

独立服务开关设在轿厢操纵盘的专用盒内。

①如轿厢正在运行期间，接通独立服务开关，轿厢将进入独立服务操作，停在最近层。

②独立服务期间，外呼信号不登记。

③如轿厢在停车期间进入独立服务，消除呼叫信号，开门。

④独立服务期间，操纵方向按钮定向且关门。若在关门过程中释放按钮，则门将重新开启。

⑤独立服务期间，关门按钮不起作用。

⑥独立服务期间，防犯罪功能仍起作用。

6）防犯罪服务

①如果接通防犯罪服务开关，轿厢将层层停车，直到到达最后一个内选楼层为止。

②断开防犯罪服务开关，防犯罪操作即刻结束。

③轿厢在运行期间，接通防犯罪服务开关，轿厢立即进入防犯罪服务操作。

④在返基站操作期间，该功能仍然有效，但轿厢到达驻停站后，该功能失效。

⑤独立服务期间，防犯罪功能仍有效。

7）检修操作

根据电梯安全规范，为便于检修和维护，应在轿顶装一个易于接收信号的控制装置，即轿顶检修盒。大多数电梯在机房和轿顶各设一个检修盒。检修开关是双稳态电气安全开关，分别可处于检修位置和正常运行位置，并设有误操作保护。

①轿顶检修优先于机房检修，即当轿顶检修开关打到检修位置，机房检修开关将不起作用。

②只有轿顶检修开关处于正常运行位置时，机房检修开关才起作用。

③检修开关一旦打到检修位置，电梯即进入检修工作状态，所有的正常运行将取消，包括任何自动门操作。只有当检修开关再次打到正常运行位置时，检修操作才结束。

④电梯正在运行时，将检修开关打到检修位置，轿厢急停并转入检修状态。

⑤检修期间，电梯运行依靠标有上、下运行方向的按钮点动运行；按下按钮，电梯即以检修速度运行，一旦松开按钮，电梯立即停车。

⑥电梯检修运行时，所有安全装置应正常有效。

8）救援操作

电梯在正常运行状态下突然停电，此时正处于平层区以外，当供电系统正常以后，控制系统会使电梯自动转入慢速返平层状态，使电梯低速运行到平层位置开门后重新恢复正常状态。

9）电梯的再平层

①如果电梯在门区内超过平层±15 mm，而且收到了控制系统发出的再平层信号，则轿厢进行再平层操作。

②若轿厢超过平层 15 mm，则向下进行再平层；若轿厢低于平层 15 mm，则向上进行

再平层。

③当门全部开启时，轿厢开始进行再平层操作，当再平层信号消失后，再平层结束。

10）电梯报警

当电梯发生故障或遇到紧急情况时，乘客可通过设置在轿厢操纵盘上的警铃按钮或其他报警装置通知值班人员，方便救护。

11）照明电路

电梯的照明电路与动力电路在控制上分开设置，目的是在动力电源出现故障时，照明仍可继续供电，以便救护乘客及方便检修。

12）消防服务

消防服务开关一般设在基站大厅电梯钥匙开关旁边，平时消防开关用玻璃板封闭，不能随意按动，而在建筑物失火时用硬器打碎面板，按下消防开关，电梯即进入消防状态。

（1）消防返基站

①消除所有轿内指令、厅外呼梯。

②若轿厢正在向基站方向运行，则轿厢将直接运行到基站，中间不停梯。

③若轿厢正在向与基站相反方向运行，则轿厢将在就近层停车，不开门，然后直接运行到基站。

④若轿厢正在某层停车开着门，则立即关门后返基站；若轿厢正在某层停车关着门，则立即返基站。

⑤轿厢运行到基站后，开门，一定时间延时后将门关闭。

⑥若轿厢已在基站待机，则立即进入消防员专用状态。

（2）消防员专用状态。

①厅外呼梯仍然消除。

②恢复轿内按钮指令功能，以便消防员操作。

③消防操作期间，开门按钮仍起作用。

④切除自动返基站功能。

13）防捣乱服务

①当轿厢的负载小于 80 kg，而内选信号在 4 个以下时，轿厢首先运行到一层，开门，消除全部信号。

②当内选信号在 4 个以上时，将消除全部内选信号。

③独立服务期间，此功能不起作用。

14）电梯的开关门

对于客梯通常要求具备自动开关门功能。

（1）电梯平层后自动开门，在门完全打开并经过一段时间延时后自动关门，延时时间可调。

（2）电梯操纵盘设有开、关门按钮，可在轿厢内手动按此按钮进行开、关门。

（3）当已定向还需继续运行的轿厢停在某层时，可通过按与运行方向相同的大厅呼梯按钮将门打开。

（4）若轿厢停在某层并处于待梯状态，则可通过大厅呼梯按钮将门打开。

（5）若门在关闭途中有障碍物碰到门安全触板（门小扇）或沐浴门扇边缘的光幕（光电保护装置），则门重新开启。

（6）当门在关闭过程中，按开门按钮，则门重新打开。

（7）轿厢超载时，超载指示灯亮，蜂鸣器发声，电梯不关门或正在关门时将重新开门，直到解除超载恢复正常为止。

单元 4　机械原理与机械设计

设计实质上是一个决策过程。假如我们有了问题，则需要设计一个解决方案。换句话说，设计就是形成一个能满足特定需求并产生一个可以实现某种实物的方案。一个糟糕的决策会产生一个糟糕的设计和产品。

在实施一个设计任务时要考虑很多因素。在许多场合下，通过常识就能解决问题。下面是一些经常要考虑的因素。

（1）采用什么装置或机构。透彻理解所要解决的问题，就可得到最好的决定。有时候一个功能可以用多种方法或不同机构来实现，设计者要选择适合该环境下最有效的方案。

（2）材料。这对任何设计都极为重要。选材错误会导致产品失败，如产品尺寸过大或过小，或者成本过高。对每一个部件来说，材料的选择取决于材料的性能、材料的可制造性及成本。

（3）负载。在零件中外部负载产生内部应力。必须精确计算这些应力，它们将决定零件尺寸。

（4）尺寸、形状、占用空间和质量。初步的分析计算可以得出零件大致的尺寸，对于标准件，就应该选取与标准尺寸相吻合的较大值。标准件的形状是已知的，非标件的形状必须由机器内可用空间来确定。按比例画一张设计图对生成初始形状和尺寸往往是很有用的。此外，质量对具体应用而言也很重要。

（5）制造。必须永远牢记，所设计的零件利用现有的设备可以容易地制造出来，且成本不高。

（6）如何操作。在设计的最后一个阶段，设计者必须确保机器操作简单。对于多数需要动力的机器而言，不过是简单地按下一个操作把柄或开关就能发动机器。而其他很多场合则需要一连串的顺序操作，这时操作顺序不能过分复杂，而且操作不应需要很大的力。

（7）可靠性与安全性。可靠性对任何设计都是一个重要的因素。所设计的机器应该高效可靠地运转。可靠性是指机器或零件在使用时不失败的概率，为 0～1（$0<R<1$）。为满足此条件，每个细节都必须检查，必须避免可能的过载、零件磨损、产生过量的热及其他有害因素。关于这一点没有单一答案，但如果总体上的安全设计方法及设计的每一个阶段都注意到了的话，即可生产出可靠的机器。

当今安全性已经成为设计中至关重要的要素。设计机器是用来服务人类而不是伤害人类的。工业法规保证机器制造商对任何因产品缺陷造成的事故负责。

（8）维修、成本及审美——维修与安全通常互相联系。好的维修可确保机器良好的运行状态。通常应有定期的维修计划，对运动和承载的零件进行彻底的检查，以免灾难性的失效发生。良好的润滑可保持较低的摩擦与磨损，这是设计过程中的主要任务。因为只要存在运动部件，摩擦与磨损就不可避免，大的摩擦将导致能量损失增加，而零件磨损将导致材料损失造成过早失效。

成本与审美是产品设计的最本质的考虑因素。成本最终取决于材料的选择，而材料的选择又取决于给定条件下的应力状况。虽然很多场合下审美考虑不是机器设计必不可少的，但必须考虑人体工程学的因素。

单元 5　零件制作与普通加工

课文 1

1. 齿轮

齿轮是直接接触、成对工作的实体。在称为齿的凸出物的连续啮合作用下，齿轮能将运动和力从一个旋转轴传递到另一个旋转轴，或从一个轴传递到一个滑块（齿条）。

（1）齿形轮廓。齿轮的接触面必须以一定的方向排齐，这样可以使得传动是正向的，也就是传递的载荷不必依靠表面的摩擦作用进行传递。像处理直接接触的实体一样，要求垂直于表面的公法线不必经过主动轴或从动轴的轴线。

与我们所知道的直接接触的实体一样，摆线和渐开线轮廓也都提供了一个正方向的驱动和一个均匀的速度比，即共轭作用。

（2）基本关系式。一对齿轮中较小的齿轮称为小齿轮，较大的齿轮称为大齿轮，若小齿轮安装在传动轴上，则这对齿轮用作减速器；反之，若大齿轮安装在传动轴上，则这对齿轮用作加速器。齿轮经常用于减速而不是加速。

如果齿轮有 N 个齿，并以 n/min 转的速度旋转，乘积 $N \times n$ 表示的是每分钟旋转的齿数。如果每个齿都是通过啮合作用传动另一个齿轮，那么这个乘积对于一对啮合齿轮的两个齿轮来说是相等的。

对于各种不同类型的共轭齿轮，齿轮比和速度比都可以通过大齿轮和小齿轮上的齿数比获得。如果一个大齿轮的齿数为 100，小齿轮的齿数为 20，则齿数比为 100 / 20=5。这样不管大齿轮的旋转速度为多少，小齿轮的旋转速度总是大齿轮旋转速度的 5 倍。大小齿轮的接触点称为节点，由于节点位于中心线上，因此节点是齿形轮廓线上唯一做纯滚动接触的点。非平行的齿轮、非交叉传动的轴也有节圆，但是不存在纯滚动节圆的概念。

齿轮的类型在很大程度上由轴的不同排列决定。一方面，即使速度发生改变，原有类型的齿轮传动一般来说也肯定好于采用其他类型的齿轮传动，这就意味着当轴的排列布置形式决定后，齿轮的类型或多或少也就定下来了；另一方面，如果齿轮的速度变化及类型一定的话，那么轴的排列布置形式基本上也就定下来了。

（3）直齿轮和螺旋齿轮。齿形轮廓是直的并且平行于传动轴的齿轮称为直齿轮。直齿轮只能用于连接平行轴。如果一个渐开线直齿小齿轮是用橡皮制成的，能均匀扭曲，则其一端会以另一端为轴进行旋转，这样小齿轮上的齿开始将是直的并行于传动轴，最后会变成螺旋形。

（4）蜗杆和伞形齿轮。为了得到线接触和改进螺旋齿轮横向轴向的传动载荷能力，部分大齿轮可能被做成弯曲形状以围绕小齿轮，有时类似于螺帽套在螺钉上。这样的结果就是圆柱形的蜗轮蜗杆。蜗杆有时也做成沙漏形状，而非圆柱状，目的是使蜗杆部分接触蜗轮，结果可以进一步提高承载能力。

蜗轮蜗杆是用一对齿轮就可以提供较大速度比的最简单方法，但是由于沿着齿面方向存在滑动现象，蜗轮蜗杆的传动总是比平行轴传动的齿轮效率更低。

2. V 形带

人造纤维和橡胶 V 形带广泛应用于动力传送。V 形带一般做成两个系列：标准 V 形带和重型 V 形带。V 形带能用于传动的中心距离较短的场合，能做成无缝的，因此避免了连接设备的麻烦。

（1）V 形带成本低，通过并排增加 V 形带的数量可以增加传动的功率，传动中的所有 V 形带被拉伸相同长度的目的是保持每条 V 形带中载荷均匀。当一条 V 形带断裂时，通常所有 V 形带都被调换。带轮可以是从上到下以一个任意角度传动的。由于 V 形带工作在一个相对小的带轮中，因此采用一个带轮就可以进行较大的减速。

（2）带轮槽的倾斜角度通常为 34°~38°。V 形带在槽中的嵌入作用可以大大增加 V 形带的牵引力。

（3）带轮可以用铸铁、钢、锻压合金件制成。在带轮槽的底部需要留有足够的间隙，以保证 V 形带不接触带轮槽的底部，因为那样的话容易磨损。有时，较大的带轮没有轮槽，此时靠带的内表面来获得牵引力，从而可以使带轮上加工槽的成本省下来。带轮运行一定时间后槽的宽度可以调节，这样带轮的有效节圆直径是变化的，因此需要进行适当的调整，以满足速度比的要求。

课文 2

车床通常被认为是最古老的机床。尽管木工车床早在 1000BC — 1AD 时就发展起来了，但带有导螺杆的金属加工车床直到 17 世纪末才发展起来。最普通的车床最初被称作普通车床，由发动机通过架空滑轮和皮带提供能量。现在这些车床都装有独立的电动机。

虽然简单且用途多，但是普通机床要求熟练的机工来操作，因为所有的控制都是手动操作的。因此，对于重复性的操作和大型流水线作业，普通机床的效率很低。

1. 机床组成

机床配有各种部件和附件。普通机床的基本部件描述如下。

（1）底座。底座支撑着机床上的所有主要部件。底座庞大且坚固，通常用灰口铁或球墨铸铁建成。底座的顶端两个部分，在使用时可强化各个横截面、加工耐磨性和尺寸精度。

（2）支撑架。支撑架或支撑架组件，沿着轨道滑行，由一个横向滑移配件、刀架和挡板组成。刀具安装在刀架上，通常采用旋转复合刀架，以便刀具定位和调整。横向滑板可以径向移进移出，以控制切削加工中（如车端面）刀具的径向位置。挡板的安装机理是使支撑架能手动和自动运行，能依靠导螺杆横向滑动。

（3）主轴箱。主轴箱固定在底座上，并配有电动机、滑轮和 V 形皮带，为在不同转速下运转的转轴提供能量。转速通过手动控制选速器来设定。绝大多数的主轴箱都装配有一组齿轮，有些主轴箱有不同驱动装置令转轴的速度可以不断变化。主轴箱有一个空心轴，工件装置（例如卡盘和钳夹）都装在上面，长棒材或管材能从中进给，完成各种车削操作。

（4）尾座。尾座能够沿着轨道滑行，并且可以在任何位置被钳住，支撑工件的另一端，它有一个可以被固定（死点）或可以随工件自由旋转（活点）的中心点。钻头和铰刀可被固定在尾座顶尖套筒（带有一个锥形孔的空心圆柱体）上，并在工件上钻轴向孔。

（5）进给杆和导螺杆。进给杆是通过主轴箱的一组滑轮来驱动的。进给杆在车床操作时可以旋转，然后靠齿轮、摩擦离合器和长杆键槽给机架和横向滑板提供运动量。闭合导螺杆周围的对开螺母使之与车架啮合，同时它也可用来精确地车螺纹。

2. 车床规格

车床通常被注明以下参数。

（1）它的摆程（所谓摆程就是车床所能加工的零件的最大直径）；

（2）主轴箱和尾座中心之间最大距离；

（3）机床底座的长度。

例如，车床可能有这些尺寸：摆程为 360 mm（14 in）、中心距离为 760 mm（30 in）、底座长度为 1 830 mm（6 ft）。一般有多种不同结构和功率的车床。

台式车床放在工作台上，它们功率低，经常用手操作进刀，用来精密加工小工件。工具车床精度高，使零件在加工时缩紧公差。普通车床的尺寸范围较广，用于各种车削加工。对于槽型机座车床，主轴箱前面的底座部分拆卸后可容纳大型直径的工件。

专用车床主要应用于铁路车轮、炮筒和轧机辊轧。可达到的加工工件的尺寸：直径为 17 m、长为 8 m（66 in×25ft），功率可达 450 kW（600 hp）。普通车床包括从 2 000 美元的台式机到 100 000 美元以上的大型设备。

3. 车床操作

在典型的车削操作中，任何一个工件夹具都可以把工件夹住。长而细的零件必须有一个稳定的支撑架支撑着并且随支撑架放到机床上，否则零件就会在切削力的作用下旋转。这些支架通常配备 3 个可调整的抓手或轧辊，用来支撑自由旋转式的工件。稳定支架直接装夹在车床的轨道上，而随行刀架装夹在车架上并随之运转。

由导螺杆来驱动的切削工具依附在刀架上，运送材料时，沿着底座运行。右切削工具朝主轴箱运动，左切削工具朝尾座运动。正面操作通过径向移动和横向滑行，夹住车架，使得到的尺寸更精确。

成形刀用于生产各种圆形的车削工件。成形刀向内径方向移动加工零件。成形刀不适合深且窄的凹槽加工或是锐角转角加工，由于它们加工时会振动，从而导致完成的表面质量较差。通常：①成形长度不能超过工件最小直径的 2.5 倍；②切削速度必须小于设置的切削速度；③必须用切削液。

车床上的镗削和车削相似。镗削在工件孔里面或钻孔里面进行。形状不规则的孔可通过镗削来矫正。工件一般装在卡盘里或采用其他合适的工件夹具装置。车床钻孔时，将钻头安装在尾架顶尖套筒（管状轴）的转头卡盘里。加工件用工件夹具固定在主轴箱上，尾架顶尖套筒靠手轮旋转向前运动。由于钻头有径向漂移的倾向，故用这种方式钻的孔可能不同轴，那么通过随后的镗削可提高钻孔的同轴性。钻过的孔以类似钻孔的方式铰大，这样可以提高孔的公差。

单元6 液压与气动回路设计

液压（又名水力学）通常指由流动液体产生的动力。现代液压的定义是使用密封的液体来传递动力、增加作用力，或者产生运动。虽然使用水车和其他简单机械产生的水力已经使用了几个世纪，但直到17世纪，液压原理才形成了科学定律。就是在那时，法国哲学家巴斯卡发现液体是不能被压缩的。他发现的定律是施加在密封液体上的压力沿所有方向传递，在相等面积上的力相等。

液压系统包含以下关键组成部分。

（1）流体——可以是任何液体。最常见的液压流体包括特制的石油炼制油，这种油对系统起到润滑作用并保护其不受腐蚀。

（2）储液罐或油箱——其作用相当于一个储存流体的仓库和散热器。

（3）液压泵——将机械能转换成液压能，通过对液体施加压力，液体从油箱流到系统里。

（4）油管——经液压系统将流体输入和输出油泵。油管可以是刚性金属管，也可以是柔性软管。油管可以在正压或负压（吸）下输送流体。

（5）液压阀——控制流体的压力、方向和流速。

（6）执行机构——将液压能转换成机械能而做功。执行机构经常采用油缸的形式。油缸应用于农业、建筑业及其他工业装备上。

在实际液压系统中，帕斯卡定律定义了从系统中产生任何结果的理论基础。这样，液压泵使系统中的流体流动。泵的进油口与油箱相连，大气压力作用于油箱内的液体上，迫使液体进入泵体。当泵运转时，其驱动液体从油箱以适当压力流进卸压管，由阀门控制受压液体的流动。在大多数液压系统中通常采用三种控制系统：①液体压力控制系统、②液体流速控制系统、③液体流向控制系统。

在这样几种情况下优先采用液压系统：①当两点之间传输动力的距离太远而不能采用带或链传动时；②当要求高扭矩、低转速时；③要求单元非常紧凑时；④当要求传动平稳无振动时；⑤速度大小及方向的控制非常简便时；⑥要求输出速度能无级调速时。

图6.1给出了电机驱动示意图。由电气驱动的油泵供有传递能量用的油量，并可传递给液压电动机或油缸，从而将液压能转换成机械能。通过阀门控制油的流动，压力油流产生线性或旋转的机械运动。油流的动能相对较低，因此有时被称为静压驱动。液压电动机和液压油缸之间几乎不存在构造上的不同。任一油泵可以被用作液压电机。在任一时间内的油流量可以通过调节阀门或采用变量泵来改变。

液压机床驱动具有众多优点，其中一个是液压驱动可在广泛的范围内提供无限变化的速度。另外，它们能像改变速度一样容易地改变驱动的方向。像许多其他类型的机床一样，许多复杂的机械装置都能够被简单化或者由于液压驱动的使用而被取消。

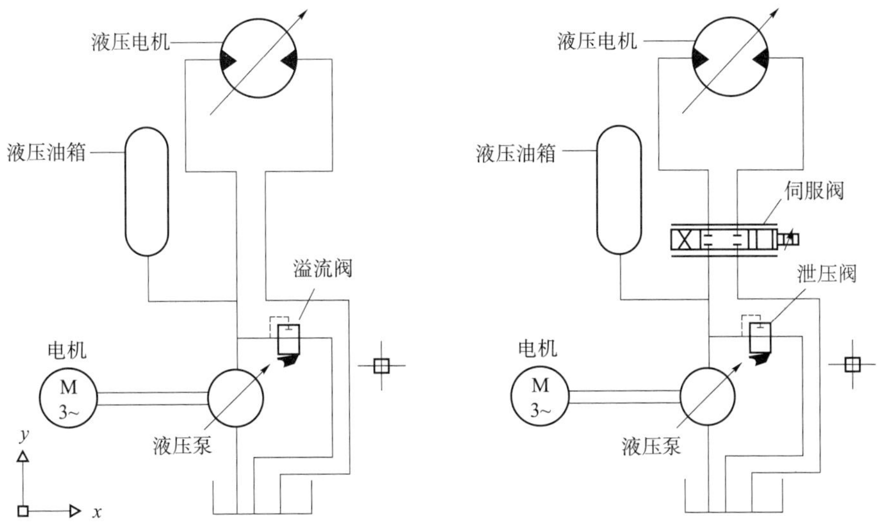

图 6.1　液压电机驱动系统

液压驱动的另一个优点是它的柔性和缓冲性。除了运行平稳外，通常可见工件的表面粗糙度得到大幅改善，刀具进刀量可以加大而不致损伤，并且持续更长时间而无须重磨。

单元 7 机械制造工艺及仿真

课文 1

为使制造过程的效率较高，制造过程中的各项活动都必须严格设计。传统上这些活动都是由工序规划师来完成的。工序设计主要关注生产方式的选择：加工、工件夹具、机器设备、操作顺序和装配。

工序和操作顺序的执行、机器设备的使用、每个操作的标准时间以及其他类似的信息都列在工序流程单表（表 7.1）。如果采用手工操作，整个任务需要大量的劳动力和时间，且在很大程度上依赖于程序规划师的经验。在工序流程单上的一个当前趋势是在计算机上存储相关数据并在零件上标上条形码（为了与其他零件区分开来）。这些产品数据可以在专用监视器上查看。

表 7.1 一个简单的工序流程单

Routine Sheet（工序流程单）

Customer's Name：Midwest Valve Co.　　Part Name：Valve Body

客户名称：中西部阀门有限公司　　零件名称：阀体

Quantity：15　　Part No.：302

数量：15　　零件号码：302

Operation No. 操作序号	Description of operation 操作描述	Machine 机器设备
10	Inspect forging, check hardness 检查锻件，校对硬度	Rockwell tester 洛氏硬度测试器
20	Rough machine flanges 粗车法兰	Lathe No. 5 五号车床
30	Finish machine flanges 精车法兰	Lathe No. 5 五号车床
40	Bore and counter bore hole 镗孔并测量镗孔	Boring mill No. 1 1 号镗床
50	Turn internal grooves 镗内部凹槽	Boring mill No. 1 1 号镗床
60	Drill and tap holes 钻孔并攻丝	Drill press No. 2 2 号钻床
70	Grind flange and end faces 刨法兰和断面	Grinder No. 2 2 号磨床

续表

Operation No. 操作序号	Description of operation 操作描述	Machine 机器设备
80	Grind bore 研磨镗孔	Internal grinder No. 1 1 号内圆磨床
90	Clean 清洁	Vapor degreaser 蒸汽去污器
100	Inspect 检查	Ultrasonic tester 超声波检验器

计算机辅助工艺设计（CAPP）是将全部工艺过程视为一个集成系统，在此基础上来完成工艺规划这个复杂的任务，这就使制造每一个零件时所涉及的单个操作和步骤能够与其他操作和步骤相协调，并且能被高效和可靠地完成。因此，计算机辅助工艺设计是CAD/CAM 的重要组成部分。

虽然 CAPP 需要应用大量的软件并且要很好地与 CAD/CAM 协调，但是它仍然是制造操作高效规划和时序安排的强有力的工具。CAPP 在小体积、高层次变体的零件产品的机加工、锻造以及类似的操作中特别有效。

1. CAPP 系统的元素

计算机辅助工艺设计有两种类型：变体式和再生式工艺设计。

在变体式系统（也称为派生系统）中，计算机文档包括为将要加工的零件制订的标准工艺计划。标准工艺计划的搜索通过对应数据库中的零件编码来实现。工艺计划的制订是依据零件的形状和加工特性来确定的。重新调整的标准工艺计划，应展示出来以供评审，并打印出来作为工艺流程单。

工艺计划包括的信息有加工过程中用到的工具和机器设备、各个加工工艺的顺序、加工速度、加工进给量、每个步骤的时间要求等。对原有工艺计划（通常是必需的计划）的小幅修改也可以实现。如果计算机文档中没有某个特殊零件的标准工艺计划，则可调出一个与其相近的工艺计划，后者与前者具有相似编码，且存有工艺单。若一个工艺单不存在，则需为这个零件制订一个并存储在计算机存储器中。

在再生系统中，工艺计划会按照传统工艺员在制订特殊零件工艺时所遵循的相同逻辑程序而自动生成。再生系统很复杂，因为系统必须包含有关零件形状和尺寸、工艺能力、加工方法的选择、所采用的机械装置、加工工具以及各个工序顺序的全面而详细的信息。

再生系统可以新建一个新的计划而不用使用和修改原有工艺计划。虽然目前它比其他系统应用得少，但是这个系统有这些优点：①适应性和一致性适合于新零件的工艺设计，增强了为新零件制订工艺计划时的灵活性和连贯性；②更高的整体设计品质，由于系统中的决策逻辑性能可以优化设计和利用最新的制造技术。

计算机的工艺设计能力可以被纳入生产系统的规划和控制之内，这些都属于计算机辅助制造的子系统。

2. CAPP 系统的优势

与传统工艺设计方法相比，CAPP 系统的优势包括以下几点。

（1）工艺设计的标准化提高了工艺设计人员的生产效率，压缩了生产周期，降低了生产成本，并提高了产品质量和可靠性。

（2）可以为具有相似形状和特征的零件准备工艺计划，它们在制造新零件时就可以很容易地搜索到。

（3）工艺计划可以被修改，以适应特殊的需求。

（4）工艺流程单可以准备得更快。与传统手写工艺流程单相比，计算机打印出来的工艺流程单更加简单明了。

（5）其他的作用，如成本预算和操作规范，都可以纳入 CAPP。

课文 2

柔性制造系统是把制造中的所有主要元件集合成为一个高自动化的系统。其首次应用于 20 世纪 60 年代末期，由一系列的制造单元组成，每个单元包含一个工业机器人（服务于多个 CNC 系统）和一个自动物料处理系统。这些都由一台中央计算机控制，对于制造过程中的不同计算机指令，它们可以下载并通过工作站依次传输给零件。

整个系统自动化程度很高，能优化整个制造过程中的每一步。这些步骤可能包括一个或多个程序和操作（如加工、磨削、切削、成形、粉末冶金、热处理和修整），还有原材料处理、检查和汇编。到目前为止，FMS 最常见的应用就是在加工和装配操作中，从机床制造中可以获得许多 FMS 技术。

在制造行业中，柔性制造系统代表着高效、高精度、高生产力，FMS 的柔性体现在它能处理各种不同的外形轮廓及以任意的顺序加工。

FMS 可以说是包含了另外两个系统的优点：①高生产率但固定传输的生产线；②可以用独立机器生产出大量多样化产品，但加工生产车间效率低。

柔性制造系统的基本元素：①工作台；②原材料和部件的自动处理和输送；③控制系统，通过柔性制造系统，物料、部件、产品进行有序流动，在生产过程中使工作台的工作效率最大。

车间内的机器类型取决于产品的类型。对于机械加工操作，它们由多个三轴或五轴加工中心、CNC 车削、铣削加工、钻孔和磨削组成。同样还包括其他各类设备，比如自动检测、装配和清洗。

其他适合 FMS 的操作包括板材成形、冲压和剪裁、锻造。它们把熔炉、锻造机器、剪裁、热处理设备和清洗设备融于一体。

由于柔性制造系统的柔性，物料处理、存储和回收等系统显得很重要。物料处理由一台中央计算机控制并由自动引导车、传送器和不同的传送机制执行。这个系统能够在任何时间、向任何机器传输不同阶段加工完成的原材料、数据块及零件。棱柱型部件通常在专门设计的托盘上传送。具有旋转对称的部件（比如用于车削加工的部件）通常由机械装置和机器人来传送。

FMS 的计算机控制系统相当于它的大脑，它包括许多硬件和软件，子系统控制车间的机器设备和原材料、数据块、零件等在加工的不同阶段从一台机器到另一台机器的传送。它同样可储存数据及提供可视化数据的通信终端。

因为 FMS 涉及巨大的资金投入，高效的机器利用率是问题的本质：机器绝对不能闲置。此外，适当的计划和过程安排很关键。

FMS 的计划安排是动态的。不像一般的车间，在那里遵循一个比较相对刚性的安排去执行一系列的操作，FMS 的日程安排计划系统具体说明了用哪种类型的操作去执行每一个零件，它指定了需要用到的机器和制造单元。动态的计划安排能够对产品类型的迅速变化做出反应，并且对实时决策反应也是灵敏的。

因为 FMS 的柔性，在制造过程中的转化可以省去安装时间，这个系统能在不同的机器以不同的顺序执行不同的操作。但是，必须对系统中的每个制造单元的特点、性能和可靠性进行检验，确保在工作站之间流动的工件满足质量及尺寸精度方面的要求。

FMS 装置都是资金密集型的，一般起价都超过 100 万美元，那么，在做决定之前必须做出全面的代价-回报分析。这个分析应该包括这些因素：资金、能源、材料和劳动力的投入，即将投产产品的市场预期，以及在市场需求和产品种类方面可能出现的波动。另外一个因素是花费在安装系统和排除系统故障上面的时间和精力。

一般地，一个 FMS 系统需要 2~5 年时间安装和至少 6 个月排除故障，尽管 FMS 需要很少的机器操作，但是负责整个操作的工作人员必须经过培训且有很高的技术。这些人员包括制造工程师、计算机程序员和维修工程师。

比起常规的制造系统，FMS 的优点如下。

（1）零件可以随意加工，不管是批量还是单件生产，并且花费少。

（2）直接的劳动和存储可以省去，比起常规系统节约了许多。

（3）对产品更改所需的准备时间减短了。

（4）产品更加可靠，因为系统有自动纠错能力，所以产品质量都是统一的。

单元 8 数控机床及应用

自动化制造领域的一个最重要的发展就是数字化控制（NC），电子工业协会给数字控制的定义是在某一点通过直接的数据插入控制运动的系统，而且该系统必须起码能自动识别数据的一部分。能制造一个部件的数据叫作部件程序，这个程序是能被控制系统读出的一系列指令，并且转化为驱动运动件的信号。过去数控加工主要集中在大批量复杂部件的制造，但是，随着更加高效程序语言的发展，现在数控加工也可用于小批量生产。数控机床包括机器控制单元（MCU）。MCU 可以再分为两部分：数据处理单元（DPU）和控制环单元（CLU）。DPU 处理从数据库传来的编码数据，并且传达每根轴上的位置信息、移动方向、反馈和辅助控制信号给 CLU，CLU 控制每根轴的实际位置和速度，并显示操作的完成时间。

数控机床的运动控制是通过把数控编码转化为机器指令而完成的，数控编码大致可以分为以下两大类。

（1）控制各个机器零件的指令，如电机启动和停止控制（将脉冲电信号输送给传达系统和逻辑控制网络就可以完成这项任务）。

（2）控制机床和工件相对运动的指令。这些指令包含一些信息，比如轴和每一个时间单元内移动的距离，它们被转化为能被机电控制系统执行的、可加工移动的控制指令。

目前，数字控制器都是通过计算机技术建立起来的，这种控制器被称为 CNC（计算机数字控制）。CNC 系统比 NC 系统柔性化程度更高，因为它可以编辑和储存程序，并且这些程序可以随时立即被调出。通常 CNC 机床比 NC 机床更容易加工复杂形状。CNC 控制器几乎可以被应用于各类机床，如车床、车铣床、钻床、磨床等。

在数字控制中，关于加工操作的每一方面的数据，如位置、速度、进给和切割液，这些都储存在一个磁盘、软盘或计算机硬盘中。数控的概念就是信息可以从存储设备中输送到机床的控制单元，基于输入信息，驱动传输装置和其他装置（硬件控制）就可以获得所需要的加工安装，比如车削有各种轮廓线的外圆这样的复杂动作都可以由它完成。如果用一台计算机给一系列的数控车床提供程序，这个系统就叫作直接的数控系统（DNC），这台计算机也同样可以给材料处理系统设备提供指令。

NC 系统控制的伺服驱动系统的每一单元都可以实现许多具体功能。一个运行的平板键盘可以允许指令进入机器，一个解码器接收从计算机来的数据并将其分为两部分：一部分为几何数据；另一部分为处理数据（包括进给速率、轴向速度和其他加工参数的信息）。几何数据包含工件移动的信息，同一组数据可用于确定在此过程的刀具长度、刀具半径、刀具补偿等；处理数据包括调整进给速度的调节功能、轴向速度、刀具改变、切割液应用等。调节功能是由进入某一表面单元的调节指令控制的，在那里这些指令与从机床传来的反馈信号对比，然后转化为特定装置的合适的控制信号。另外，一个连接装置可以作为防范危险的安全开关。当出现信号冲突时停止机器，以防损坏机器或导致工伤。几何数据只能在合适的工件-机床关系调整之后才能使用，这个“适当”允许程序员在不考虑工件实际位置的情况下编辑程序。正确的计算可以考虑在机床几何尺寸的情况下进行，比如钻头长度、车刀尺寸、铣刀半径等的计算。

单元 9　PLC 控制技术与应用

1. PLC 是什么

可编程逻辑控制器（PLC）是一种为取代机器控制中必须用到的顺序继电器电路而发明的装置。PLC 工作时，首先检测其输入，根据其输入的状态由程序决定其输出的接通或断开。程序通常由用户用软件输入，它给出用户所期望的结果。

PLC 用于许多现实应用中，哪里有工业，哪里就可能有 PLC。如果你正在从事机械加工、包装、材料处理、自动装配或其他行业，你可能已经用到了 PLC。若还没有，那么你正在浪费时间和金钱。几乎所有需要某种电气控制的应用都需要用到 PLC。

例如，假设接通开关后，我们想让一个电磁铁接通 5 s 后断开，不考虑断开多长时间。这样的问题用一个外部定时器就可以解决。但如果这一过程含有 10 个开关和 10 个电磁铁，我们就需要使用 10 个外部定时器。假如该过程还需要计算开关分别接通的次数，那么就需要使用很多的外部计数器。如你所见，这个过程越是庞大，就越是需要使用 PLC。对于上述问题，我们可以简单地编写一个程序，对其输入进行计数，接通电磁铁一直到指定的时间。

2. PLC 的历史

PLC 最早于 20 世纪 60 年代末期出现。设计此装置的主要目的是消除因采用基于机器控制的复杂继电器带来的巨大成本。位于马萨诸塞州的 Bedford 公司给美国一家著名的汽车制造商提供了一种叫作模块化数字控制（MODICON）的方案，在当时还有另外一些公司提出了基于 PDP-8 的方案。Modicon 084 是世界上第一台用于生产的商业 PLC。

当生产需求变化时，控制系统也要随之变更。当变化频繁时，就显得十分昂贵。由于继电器是机器装置，故具有一定的使用寿命限制，且要求严格进行定期维护。当继电器数量众多时，查错纠错也非常琐碎。设想一个含有几百个甚至几千个继电器的控制面板，其尺寸大得吓人。如此多的装置，其最初的接线该有多么复杂。为了达到期望的结果，要把这些继电器一个一个地连线。这样会有问题吗？当然有！

这些“新控制器”必须容易编程和维护，寿命必须足够长而且编程改变容易实现，其还必须在严酷的环境下运行良好。那么多问题到底如何解决？答案是采用大多数人都已经很熟悉的编程技术，并且用固态继电器替换机器零件。

在 20 世纪 70 年代中期，主要的 PLC 技术是顺序状态机及基于位片的 CPU。其典型的 AMD2901/2903 在当时的 Modicon 系列及 AB 公司的 PLC 中应用相当普遍。

通信功能开始在 1973 年出现。第一台具有通信功能的 PLC 是 Modicon 系列的 Modbus。一台 PLC 可以与其他的 PLC 通信，也可以将其放在远离各自受控机器的地方。它们也能发送/接收可变电压信号，由此进入模拟世界。遗憾的是，由于缺乏标准化，在不断变化的技术领域，因协议及物理网络层互不兼容，PLC 通信最终被淘汰。尽管如此，这仍然是 PLC 辉煌的十年。

20 世纪 80 年代出现了以通用汽车公司的生产自动化协议（MAP）为通信标准化的尝试。同时，PLC 的尺寸也得到了减小，并且能够在个人计算机上通过软件进行符号编程，而不是在专用编程终端或手持设备上进行。现在，最小的 PLC 大约只有一台控制继电器般大小。

20 世纪 90 年代开始逐渐减少了新协议的引入，但是那些在 20 世纪 80 年代幸存下来的较有名协议的物理层开始了现代化进程。最新的标准（IEC 1131-3）已经试图将 PLC 编程语言标准化。我们现在的 PLC 可以用功能模块图、指令表、C 及结构化语言进行编程。个人计算机也开始取代 PLC 在某些领域应用。最早发明 PLC 的 Modicon 公司实际上已经转向基于个人计算机的控制系统了。

下一个十年将会带来什么？只有时间能回答。

单元 10 工业机器人及应用

1. 定义

“机器人是一种可以重复编程的多功能机器。它通过改变各种程序化的动作来对材料、零件、工具或专用设备进行操作，以提高任务的完成质量。”

——机器人工业协会

“机器人是一种用来执行一些通常由人完成的任务的自动化装置，或者说机器人是一种具有人形的机器。”

——韦伯大辞典

2. 历史

机器人 robot 这个词来源于捷克文，意为苦役或农奴。机器人学 robotics 指有关机器人的研究和使用。

最早的工业机器人取名为 Unimates，是 George Devol 和 Joe Engelberger 于 20 世纪 50 年代末 60 年代初开发的。此后 Engelberger 成立了 Unimation 公司，并将机器人推向市场。他也因此被称为“机器人之父”。

3. 关键组件

虽然机器人种类繁多，但它们的内部构造和组成是相同的。机器人的关键组件如下：

（1）电源转换单元：给机器人其他部件提供电源。

（2）传感器：测量机器人的状态或配置及其环境（如手臂位置、周围是否存在有毒气体），并以电信号的形式将此信息传送给机器人的控制器。

（3）执行器：利用多种机电设备（如同步电机、步进电机、交流伺服电机、直流无刷或直流有刷电机等）来完成某种任务。

（4）控制器：提供必要的智能控制。例如，处理传感器信息，对执行器控制指令进行运算，并让它们完成指定任务。

（5）用户界面：这是供用户操作的部分，即控制面板。

（6）操作基座：机器人可以在固定的基座上工作，也可以在移动的基座上工作，即机器人可以用轮子或腿移动。

4. 工业上的运用

目前 90%的机器人工作在工厂里，所以被称为工业机器人。十年前，90%的机器人被汽车制造商所购买，现在只有 50%。机器人慢慢地在仓库、实验室、研究勘探现场、能源工厂、医院甚至太空得到应用。

至少可以说，机器人工业正在迅速发展。2007 年上半年，北美地区机器人供应商的订货单飙涨了 36%，这是机器人工业协会的最新统计。

有很多理由说明机器人在工业上非常有用。安装机器人往往能使公司更具竞争力，因为机器人在做很多事时都比人更有效率。

机器人不会生病或者需要休息，它们可以一天 24 h 不间断地工作。

当某项任务由人工完成可能出现危险时，就可以考虑使用机器人。

机器人不会对工作产生厌倦，重复劳动或者无报酬对它们来说不成问题。

虽然并非所有工作都能胜任，但机器人将一些工业领域内的工作完成得更好。这些任务包括以下几项。

（1）装配——在全世界现有机器人应用中，装配作业约占33%，可以看到很多机器人被用于汽车或者电子工业的装配线上。

（2）连续弧焊与点焊——最常见的用途就是焊接。例如，用机器人焊接的汽车车体能提高安全性，因为机器人不会错过每一个焊点，而且在一天当中的每一个时刻工作得同样好。大约有25%的机器人用于各种焊接作业中。

（3）包装/堆码——包装/堆码作业仍然是工业机器人应用较少的一个领域，仅占2.8%（1997）。

在大规模工业生产中，以下工种最有可能被机器人取代：喷镀/油漆、材料清除、机器装载、材料传输、切割操作、零件检测、零件分类、零件清洗、零件抛光。随着机器人变得更加便宜，我们将看到更多的机器人接替人类来完成相应的工作。

单元 11　机械产品质量检测与装配

1. 产品质量检测

产品质量检测指的是根据产品标准或检验规程对产品的一个或多个质量特性进行试验、测量，并将结果和规定的质量要求进行比较，以确定每项质量特性合格情况的技术活动。

根据产品的使用要求不同，每种产品都有各自的质量特性。这些特性一般都转化为具体的质量要求，并在产品的技术标准（国家标准、行业标准、企业标准）和其他相关的产品设计图样、工艺制造技术文件中明确规定，以作为质量检验的依据和检验后比较检验结果的参照基准。为了保证产品质量，必须对生产过程中的原材料、外购件、外协件、毛坯、半成品、成品等进行质量检验，严格把关，使不合格的原材料不投产，不合格的半成品不转序，不合格的零件不装配，不合格的产品不出厂，保护国家和消费者利益，维护生产者信誉和提高社会效益。产品质量检测是生产中质量管理的一个重要组成部分。

2. 机械装配

机械产品一般都是由许多个零件和部件组成的。按照规定的技术要求，将若干个零件组合成组件、部件或将若干个零件的组件、部分组成产品的过程，称为装配。

机械装配是整个机械制造过程中的最后一个阶段，在制造过程中占有非常重要的地位。机械产品的质量最终由装配工作保证。零件质量是机械产品质量的基础，但装配过程并不是将合格零件简单地组合起来。即使使用了高质量的零件，低质量的装配也可能装出低质量的产品；高质量的装配则可以在经济精度零件、部件的基础上，装配出高质量的产品。

近年来，由于在毛坯制造和机械加工等方面的机械化、自动化程度提高较快，装配工作量在制造过程中所占的比重有扩大的趋势。因此，必须提高装配工作的技术水平和劳动生产率，才能适应整个机械工业的发展趋势。

对于结构比较复杂的产品，为了保证装配质量和装配效率，需要根据产品的结构特点从装配工艺角度将产品分解为单独进行装配的装配单元。

零件是组成机械产品的最基本的单元，零件一般装配成合件、组件或部件后再装配到机器上。合件也称为套件，是由若干个零件永久连接而成或连接后再经加工而成的。组件是若干个零件和合件的组合。零件在机器中能完成一定的、完整的功用。

3. 三坐标测量仪

三坐标测量仪三轴均有气源制动开关及微动装置，可实现单轴的精密传动，采用高性能数据采集系统，应用于产品设计、模具装备、齿轮测量、叶片测量、机械制造、工装夹具、电子电器等精密测量。

三坐标测量仪是指在一个六面体的空间范围内，能够表现几何形状、长度及圆周分度等测量能力的仪器，又称为三坐标测量机或三坐标量床。三坐标测量仪又可定义为“一种具有可做三个方向移动的探测器，可在三个相互垂直的导轨上移动，此探测器以接触或非接触等方式传递信号，三个轴的位移测量系统（如光栅尺）经数据处理器或计算机等计算出工件的各点（x，y，z）及各项功能测量的仪器”。三坐标测量仪的测量功能应包括尺寸精度、定位精度、几何精度及轮廓精度等，广泛地应用于汽车、电子、机械、航空、军工、模具等行业中的箱体、机架、齿轮、凸轮、蜗轮、蜗杆、叶片、曲线、曲面等的测量。

简单地说，三坐标测量仪就是在三个相互垂直的方向上有导向机构、测长元件、数显装置，有一个能够放置工件的工作台，测头可以以手动或机动的方式轻快地移动到被测点上，由读数设备和数显装置把被测点的坐标值显示出来的一种测量设备。显然这是最简单、最原始的测量机。有了这种测量机后，在测量容积里任意一点的坐标值都可通过读数装置和数显装置显示出来。测量机的采点发信装置是测头，在沿 x、y、z 三个轴的方向装有光栅尺和读数头。其测量过程就是当测头接触工件并发出采点信号时，由控制系统去采集当前机床三轴坐标相对于机床原点的坐标值，再由计算机系统对数据进行处理。

单元 12　逆向工程与 3D 打印

1. 逆向工程

逆向工程（RE），又称逆向技术，是一种产品设计技术的再现过程，即对一项目标产品进行逆向分析及研究，从而演绎并得出该产品的处理流程、组织结构、功能特性及技术规格等设计要素，以制作出功能相近，但又不完全一样的产品。逆向工程源于商业及军事领域中的硬件分析。其主要目的是在不能轻易获得必要的生产信息的情况下，直接从成品分析，推导出产品的设计原理。逆向工程的实施过程是多领域、多学科的协同过程。

在工程技术人员的一般概念中，产品设计过程是一个从设计到产品的过程，即设计人员首先在大脑中构思产品的外形、性能和大致的技术参数等，然后在详细设计阶段完成各类数据模型，最终将这个模型转入研发流程，完成产品的整个设计研发周期。这样的产品设计过程称为“正向设计”过程。逆向工程产品设计可以认为是一个从产品到设计的过程。简单地说，逆向工程产品设计就是根据已经存在的产品，反向推出产品设计数据（包括各类设计图或数据模型）的过程。从这个意义上说，逆向工程在工业设计中的应用已经很久了。比如早期的船舶工业中常用的船体放样设计就是逆向工程很好的实例。

2. 逆向工程的作用

逆向工程被广泛地应用到新产品开发和产品改型设计、产品仿制、质量分析检测等领域，它的作用如下。

（1）缩短产品的设计、开发周期，加快产品的更新换代速度。

（2）降低企业开发新产品的成本与风险。

（3）加快产品的造型和系列化的设计。

（4）适合单件、小批量的零件制造，特别是模具的制造，包括直接制模法与间接制模法。

3. 逆向扫描

所谓逆向扫描技术就是对实物原形进行 3D 扫描、数据采集，经过数据处理、三维重构等过程，构造具有相同形状结构的三维模型。逆向扫描的目的是利用实物获取点云，并基于点云进行优化设计以及创新设计。逆向扫描技术具有三维展示性，可以运用软件对物体结构进行多方位扫描，从而建立物体的三维数字模型。

逆向扫描原理：采用一种结合结构光技术、相位测量技术、3D 视觉技术的复合三维非接触式测量技术。所以又称之为“三维结构光扫描仪”。采用 3D 扫描技术，使得对物体进行照相测量成为可能。所谓照相测量，就是类似于照相机对视野内的物体进行照相，不同的是照相机摄取的是物体的二维图像，而测量仪获得的是物体的三维信息。与传统的三维扫描仪不同的是，该扫描仪能同时测量一个面。

4. 3D 打印

3D 打印（3DP）即快速成形技术的一种，又称增材制造，它是一种以数字模型文件为基础，运用粉末状金属或塑料等可黏合材料，通过逐层打印的方式来构造物体的技术。

3D 打印通常是采用数字技术材料打印机来实现的，常在模具制造、工业设计等领域用于制造模型，后逐渐用于一些产品的直接制造，已经有使用这种技术打印而成的零件。该技术在珠宝、鞋类、工业设计、建筑、工程和施工（AEC）、汽车、航空航天、牙科和医疗产业、教育、地理信息系统、土木工程、枪支以及其他领域都有所应用。

日常生活中使用的普通打印机可以打印计算机设计的平面物品，而所谓的3D打印机与普通打印机工作原理基本相同，只是打印材料有些不同，普通打印机的打印材料是墨水和纸张，而3D打印机内装有金属、陶瓷、塑料、砂等不同的“打印材料”，是实实在在的原材料。打印机与计算机连接后，通过计算机控制可以把“打印材料”一层层叠加起来，最终把计算机上的蓝图变成实物。通俗地说，3D打印机是可以“打印”出真实的3D物体的一种设备，比如打印一个机器人、打印玩具车、打印各种模型，甚至是食物，等等。之所以通俗地称其为“打印机”是参照了普通打印机的技术原理，因为分层加工的过程与喷墨打印十分相似。这项打印技术称为3D立体打印技术。

3D打印存在着许多不同的技术。它们的不同之处在于以可用材料的方式，并以不同层构建或创建部件。3D打印常用材料有尼龙玻纤、耐用性尼龙材料、石膏材料、铝材料、钛合金、不锈钢、镀银、镀金、橡胶类材料。

单元 13　智能制造生产线及应用

智能制造生产线是把制造中的所有主要元件集合成为一个高自动化的系统，首次应用于 20 世纪 60 年代末期，由一系列的制造单元组成，每个单元包含一个工业机器人（服务于多个 CNC 系统）和一个自动物料处理系统。这些都由一台中央计算机控制，对于制造过程中的不同计算机指令，它们可以下载并通过工作站依次传输给零件。

整个系统自动化程度很高，它能优化整个制造过程中的每一步。这些步骤可能包括一个或多个程序和操作（比如加工、磨削、切削、成形、粉末冶金、热处理和修整），还有原材料处理、检查和汇编。到目前为止，智能制造生产线最常应用于加工和装配操作中，从机床制造中可以获得许多的智能制造生产线技术。

在制造行业中，智能制造生产线代表着高效、高精度、高生产力。

1. 智能制造生产线单元

一个智能制造生产线的基本元素如下：

（1）工作台；

（2）原材料和部件的自动处理和输送；

（3）控制系统。通过这个系统，物料、部件、产品进行有序流动，在生产过程中工作台的工作效率最大。

车间里的机器类型取决于产品的类型。对于机械加工操作，它们由多个 3~5 轴加工中心、CNC 车床、铣削床、钻床和磨床组成。同样还包括其他各类设备，比如自动检测、装配和清洗。

其他适合智能制造生产线的操作包括板材成形、冲压和剪裁、锻造。它们把熔炉、锻造机器、剪裁、热处理设备和清洗设备融于一体。

由于智能制造生产线的柔性，物料处理、存储和回收等系统显得很重要。物料处理由一台中央计算机控制并由自动引导车、传送器和不同的传送机制执行。这个系统能够在任何时间、向任何机器传输不同阶段加工完成的原材料、数据块及零件。棱柱形部件通常在专门设计的托盘上传送。具有旋转对称的部件（比如用于车削加工的部件）通常由机械装置和机器人来传送。

智能制造生产线的计算机控制系统相当于它的大脑，它包括许多硬件和软件，子系统控制车间的机器设备和原材料、数据块、零件等在加工的不同阶段从一台机器传送到另一台机器。它同样可储存数据和提供可视化数据的通信终端。

2. 安排调度

因为智能制造生产线涉及巨大的资金投入，高效的机器利用率是问题的本质：机器绝对不能闲置。此外，适当的计划和过程安排很关键。

智能制造生产线的计划安排是动态的。不像一般的车间，在那里遵循一个比较相对刚性的安排去执行一系列的操作，智能制造生产线的日程安排计划系统具体说明了用哪种类型的操作去执行每一个零件，它指定了需要用到的机器和制造单元。动态的计划安排能够对产品类型的迅速变化做出反应，并且对实时决策反应也是灵敏的。

由于智能制造生产线的柔性，在制造过程中的转化可以省去安装时间，这个系统能在不同的机器以不同的顺序执行不同的操作。但是，必须对系统中的每个制造单元的特点、性能和可靠性进行检验，确保在工作站之间流动的工件满足质量及尺寸精度方面的要求。

3. 智能制造生产线在经济上的合理性

智能制造生产线装置都是资金密集型的，一般起价都超过一百万美元，那么，在做决定之前必须做出全面的代价-回报分析。这个分析应该包括这些因素：资金、能源、材料和劳动力的投入，即将投产产品的市场预期，以及在市场需求和产品种类方面可能出现的波动。另外一个因素是花费在安装系统和排除系统故障方面的时间和精力。

一般地，一个智能制造生产线系统需要 2 ~ 5 年时间安装和至少 6 个月排除故障，尽管智能制造生产线需要很少的机器操作，但是负责整个操作的工作人员必须经过培训且有很高的技术。这些人员包括制造工程师、计算机程序员和维修工程师。

与常规的制造系统相比，智能制造生产线的优点如下。

（1）零件可以随意加工，不管是批量还是单件生产，并且花费少。

（2）直接的劳动和存储可以省去，比起常规系统节约了许多。

（3）对产品更改所需的准备时间减短了。

（4）产品更加可靠，因为系统有自动纠错能力，所以产品质量都是统一的。

单元 14 机电设备保养与维护

1. 机电设备保养的概念

机电设备使用的前提和基础是设备的日常维护和保养，设备维护保养包含的范围较广，包括：为防止设备劣化，维持设备性能而进行的清扫、检查、润滑、紧固以及调整等日常维护保养工作；为测定设备劣化程度或性能降低程度而进行的必要检查；为恢复设备性能而进行的修理活动。

2. 机电设备保养工作的内容

对设备的保养的主要依据是各类不同设备的使用要求，遵循使用说明书有针对性地进行保养，主要包括防锈、防蚀、清洗、换油、电源维护以及其他必要的调整工作，比如调整配合间隙、紧固零件等。

防锈和防蚀是设备保养中最为复杂的工作。在生产中应该根据设备所处的环境，综合考虑温度、湿度、工业性气氛（如二氧化硫、二氧化碳等酸性气体）、粉尘等因素，选择恰当的物理防锈或电化学防锈方法对设备整体和零件加以保护。保养设备时还应注意机架和机壳的卫生，定期清理设备表面及内部的污染物，对易磨损部件及时添加油或更换零件。此外，也应定期检查各种保护、防护设施是否完好，如接地装置、防护屏障、标示牌、警示牌等。电源是所有机电设备的动力来源，设备的正常运行依赖于稳定、可靠的电源供给环境，因此，也必须经常性地对电力线路、配电柜、配电箱等电力设施进行检查，加强对供电线路的管理，防止鼠害、老化等情况的发生，发现问题应及时处理。

3. 机电设备维修保养的重要性

（1）为机电设备正常运转提供保障。

机电设备是新时期很多企业产品加工的重要保障，做好其维护保养工作，能够及时发现设备存在的问题和隐患并第一时间解决，如此可以为设备安全运行奠定良好基础，从而有效提高企业产品生产秩序。

结合大量实践来看，在企业生产加工中，一旦机电设备产生故障，不仅会影响作业效率，还容易引发各种生产事故。所以提高机电设备维修及保养重视程度至关重要，这也是提高设备应用率、保证所有功能模块和零件正常运行的关键所在。

现阶段，国内企业生产规模不断扩大，很多机电设备需要长时间投入使用，这也在一定程度上加大了内部结构和零部件的磨损率，只有根据机电设备运行情况、运行环境等要素做好维修保养工作，才能及时、准确地判断设备是否处于正常运行状态，有利于为设备安全使用、正常运转提供保障。

（2）稳定机电设备功能性。

做好机电设备维修与保养工作，在稳定设备功能方面发挥积极作用。众所周知，机电设备投入使用后，受运行环境等因素影响，容易导致部分功能受到制约，这就需要维修保养部门对机电设备本身及运行环境进行深入分析，总结出影响设备功能的各种因素，并采取有效措施解决。

在企业生产中，一旦机电设备出现功能与标准不符的情况，则大多是由设备运行参数不合理、运行程序紊乱引起的，需要及时通过维修与保养进行调整。尤其是在新形势下，企业产品加工精度越来越高，如果机电设备功能方面存在障碍，则会导致生产出的产品与精度要求产生偏差，不利于提高产品合格率，会给企业造成巨大经济损失。而定期开展机电设备维修与保养工作，可以有效地提高设备功能的稳定性和可靠性，从而保证生产任务有条不紊地进行。

4. 机电设备日常维护保养注意事项

为了提升机电设备运行稳定性，尽可能排除故障隐患，为企业生产加工提供保障，需要在日常维护保养中注意以下事项。

第一，做好机电设备日常清洁工作，保证设备内外干净、整洁，避免零件因灰尘过多或杂物堵塞出现过热、停止运行等故障。另外，在日常保养过程中，还要注意观察油孔、滑动面是否出现漏油问题，如果存在要及时找到原因并解决。在设备投入使用前，应注意检查设备表面完整性，并做好试验工作，确保运行稳定即可使用，如果出现漏气、漏油问题，要及时与厂家沟通并维修。

第二，在机电设备投入使用后，需要将其设置在合适位置，并合理规划线路、管道等，保证设备运行始终处于良好状态。

第三，时刻观察机电设备各零部件的润滑度，如果油量不足要及时补充，避免零部件因缺油出现相互摩擦的现象。在此基础上，还要注意对油压进行测量，补油时选择质量过关的产品，定期对油枪和油杯进行清理。

第四，工作人员要熟练掌握机电设备的操作方法，并严格规范自身行为，在使用过程中要注重使用与之相适应的安全防护手段，确保设备运行稳定，为生产作业有序进行奠定基础。

单元 15　生产现场优化与数字化管理

生产现场管理是企业管理的重要组成部分，管理水平的高低直接影响企业的生产效率和经济效益。优化生产现场管理是企业管理实现整体优化的前提和基础。企业的生存和发展遵循优胜劣汰的客观规律，主要取决于市场产品和管理水平两个因素。企业竞争优势的重要标志包括产品质量、服务水平、交货期以及生产成本，稳定的产品质量是企业在激烈的市场竞争中取胜的法宝。对生产现场科学、细致的管理是产品质优价廉的重要保障。

1. 生产现场优化管理简介

生产现场优化管理以提高企业的现场管理水平、提高生产现场相关人员的改善意识及技能水平为目的，以如何提高生产流动性为主线，进行改善生产现场生产效率、品质、交货期、安全、环保、员工士气等。通过对员工的生产现场管理内容的培训可以改变员工的工作理念、提高工作技能，使员工能积极地参与到现场管理中去，从而在员工自身各方面得到提高的同时，企业能得到更高的回报。

2. 生产现场管理的意义

（1）建立企业优势实力。

提高生产现场管理水平是建立企业优势实力的基础。企业优势实力是指企业在争夺市场和争夺顾客方面具有高于竞争对手的实力。企业的优势实力包括资金、人力和技术、市场营销、产品质量、产品成本等。建立以生产现场为中心的综合管理体系，是挖掘生产潜力、提高经济效益的利器。

（2）全面提高企业素质。

提高生产现场管理水平是提高企业经营管理素质、生产技术素质、基础工作素质、职工队伍素质和精神文明素质等的前提和保证，否则全面提高企业素质只能是一句空话。现场管理的好坏是衡量企业管理水平高低的重要标志。

（3）提高企业经济效益。

提高企业经济效益必须从生产现场抓起，因为企业效益来源于全体职工劳动所创造的价值，而这些价值大多实现于生产现场。只有加强生产现场管理，以最低的投入获得最高限度的产出，才能合理配置和有效利用生产要素。

3. 数字化管理

数字化生产管理是指利用计算机、通信、网络等技术，量化管理对象与管理行为，实现研发、计划、组织、生产、协调、销售、服务、创新等职能的管理活动和方法。

数字化生产管理内容如下。

①精益运营提升：以精益思想对企业经营和生产现场进行优化改善；②现场智能管理：通过工业物联网对生产现场进行数字化智能管控；③设备运维管理：通过机联网平台掌握设备实时情况实现远程监管；④质量监控管理：对生产重要环节进行管控，实现全流程可追溯管理；⑤仓储物流管理：采用扫描与感应进行数据采集实现智能仓库作业。

References

[1] 王晓江. 机械制造专业英语［M］. 北京：机械工业出版社，2017.

[2] 沈言锦，漆江艳，易剑英，等. 机电专业英语［M］. 2 版. 北京：机械工业出版社，2024.

[3] LYSHEVSKI S E. Mechatronic curriculum-retrospect and prospect［J］. Mechatronics，2012，12：195-205.

[4] JONATHAN W. An introduction to mechanical engineering［M］. John Wiley & Sons，Inc.，2006.

[5] 王鹏飞. 机电专业英语［M］. 北京：北京理工大学出版社，2019.

[6] 叶久新，童长清. 模具专业英语教材［M］. 2 版北京：北京理工大学出版社，2019.

[7] KEVIN C. PLC Programming for industrial automation［M］. Massachusetts：O'Reilly Media，2008.

[8] CRAIG J J. Introduction to robotics：Mechanics and control［M］. New York：Dover Publications，2005.

[9] MILLS D. Advances in solar thermal electricity technology［J］. Solar Energy，2004（1/3）76.

[10] 岳殿霞，张辉. 机电专业英语［M］. 北京：北京交通大学出版社，2021.

[11] 白庆华. 工业工程专业英语［M］. 上海：上海交通大学出版社，2009.

[12] 董建国. 机械专业英语（高职）［M］. 3 版. 西安：西安电子科技大学出版社，2018.

[13] 肖平，韩利敏. 机械工程专业英语［M］. 西安：西安电子科技大学出版社，2017.

[14] KALPAKJIAN S. Manufacturing processes for engineering materials［M］. 5th ed. New Jersey：Pearson Education Inc，2003.

[15] TOOGOOD R. Pro/Engineer tutorial［M］. New York：Prentice Hall，2005.

[16] 赵运才，何法江. 机电工程专业英语［M］. 北京：北京大学出版社，2006.

[17] 朱晓玲. 机电工程专业英语［M］. 北京：机械工业出版社，2007.